The Baofeng Radio Bible

Bud Douglas

© Copyright 2024 - All rights reserved.

The content contained within this book may not be reproduced, duplicated or transmitted without direct written permission from the author or the publisher.

Under no circumstances will any blame or legal responsibility be held against the publisher, or author, for any damages, reparation, or monetary loss due to the information contained within this book. Either directly or indirectly.

Legal Notice:

This book is copyright protected. This book is only for personal use. You cannot amend, distribute, sell, use, quote or paraphrase any part, or the content within this book, without the consent of the author or publisher.

Disclaimer Notice:

Please note the information contained within this document is for educational and entertainment purposes only. All effort has been executed to present accurate, up to date, and reliable, complete information. No warranties of any kind are declared or implied. Readers acknowledge that the author is not engaging in the rendering of legal, financial, medical or professional advice. The content within this book has been derived from various sources. Please consult a licensed professional before attempting any techniques outlined in this book.

By reading this document, the reader agrees that under no circumstances is the author responsible for any losses, direct or indirect, which are incurred as a result of the use of information contained within this document, including, but not limited to, — errors, omissions, or inaccuracies.

TABLE OF CONTENTS

Introduction 4

Glossary of Radio Terms 8

Chapter 1: A Beginner's Guide to Setup and Basic Functions 11

Chapter 2: Understanding Buttons, Menus, and Displays 14

Chapter 3: Frequencies, Channels, and CHIRP Basics 20

Chapter 4: Choosing the Right Tool for the Job 25

Chapter 5: Tips and Tricks for Clear Communication 30

Chapter 6: Effective Communication Techniques 35

Chapter 7: Using Your Baofeng in Critical Situations 41

Chapter 8: Digital Modes 47

Chapter 9: Troubleshooting and Maintenance 50

Chapter 10: Legal and Regulatory Landscape 54

Chapter 11: Specialized Communication Protocols 57

Chapter 12: Baofengs for Preppers and Survivalists 61

Chapter 13: Ham Radio Operations with Baofengs 65

Chapter 14: Unleashing the Potential 71

Conclusion: The Future of Baofeng and Your Communication Toolkit 77

BONUS 1: Essential Reference: Frequencies, Phonetics, and More 78

Bonus 2: Upgrade Paths for Your Baofeng 81

GET YOUR 5 BAOFENG BONUSES!

SCAN HERE TO DOWNLOAD IT

Introduction

Have you ever held a small, unassuming object in your hand, knowing it held the potential to connect you across vast distances, through dense forests, and even in the face of emergencies? That's the magic of the Baofeng radio, a two-way communication device that has captured the imaginations of hobbyists, preppers, and adventurers alike.

But for many, the Baofeng can seem shrouded in mystery. Buttons, knobs, dials, and cryptic acronyms might leave you wondering, "Where do I even begin?" This book is your key to unlocking the potential of your Baofeng radio, transforming it from a confusing gadget into a powerful tool for communication, exploration, and preparedness.

Whether you're a complete beginner or a curious tinkerer, we'll embark on a journey together. We'll demystify the technical jargon, navigate the programming process with confidence, and explore the vast capabilities of your Baofeng.

Here's a glimpse of what awaits you in this guide:

Unboxing the Basics: We'll start by getting familiar with the anatomy of your Baofeng, understanding its buttons, functions, and essential features.

Speaking the Language of Radio: No more feeling lost in a sea of abbreviations! We'll break down key radio terms like frequencies, channels, and modes, making you a fluent speaker of the radio language.

Programming Power: Programming your Baofeng unlocks its true potential. We'll guide you step-by-step through different methods, ensuring you can connect with others on your chosen frequencies.

Beyond the Basics: Once you've mastered the fundamentals, we'll delve into advanced techniques like antenna selection, signal boosting, and exploring digital modes for secure communication.

Practical Applications: From staying connected during outdoor adventures to preparing for emergencies, we'll showcase the real-world applications of your Baofeng radio, empowering you to use it in meaningful ways.

Community and Connection: The world of Baofeng enthusiasts is vast and welcoming. We'll explore how to connect with other radio users, learn from their experiences, and expand your communication network.

Remember, this journey is not just about mastering the technical aspects of your Baofeng. It's about unlocking the potential for connection, exploration, and preparedness that lies within this remarkable device. So, grab your Baofeng, turn the page, and let's begin uncovering its hidden potential, together.

The Baofeng Revolution: From Budget to Benchmark

The world of two-way radios was once dominated by complex, expensive equipment out of reach for many enthusiasts and everyday users. Baofeng shattered this paradigm, offering surprisingly feature-rich radios at prices that redefined the concept of accessibility. This disruptive force emerged from humble origins in China as a manufacturer of simple, low-cost radios. While unremarkable initially, these models found a niche and caught the attention of importers, who recognized the potential to bring affordable radio communication to a global audience.

Baofeng's surge in popularity was particularly notable within the amateur radio (ham radio) community. Licensed hams with an experimental spirit and technical know-how discovered that models like the now-iconic UV-5R could, with a bit of tinkering, offer performance rivaling far pricier brands. This fueled their widespread adoption and launched a wave of modifications and optimizations. Baofeng, however, was not without its critics. Questions surrounding their out-of-the-box compliance with varying radio regulations (particularly in the US and Europe) ignited debates and concerns about quality. These criticisms weren't ignored; they became a catalyst for Baofeng to improve both quality and adherence to specific region-based regulations.

Beyond amateur circles, Baofengs proved their worth to countless other users. Preppers saw the value of reliable communication, especially given the affordability factor. Small businesses embraced Baofengs as a cost-effective way to connect their staff. Hobbyists and experimenters found them an ideal platform for DIY projects, custom antenna builds, and even software tweaks.

The impact of Baofeng extends beyond the radios themselves. They proved quality two-way communication shouldn't be exclusively for those with deep pockets. This ignited online communities dedicated to Baofeng knowledge sharing, YouTube channels filled with tutorials, and even spurred established radio manufacturers to innovate in both features and affordability.

A Beacon of Hope in Crisis

The true impact of the Baofeng revolution became strikingly evident in the aftermath of the devastating 2010 Haiti earthquake. With traditional communication infrastructure crippled, a network of volunteer ham radio operators armed with affordable Baofeng radios stepped into the gap. They coordinated relief efforts, relayed vital medical information, and even helped reunite families separated by the chaos. This demonstrated that reliable communication empowered by even the most humble-seeming radio could literally save lives.

While today's Baofeng lineup boasts models from basic to advanced, the core philosophy of accessible radio ownership remains strong. Their journey is one of disruption, evolution, and empowering a vast range of individuals to experience the thrilling and practical world of radio communication.

Glossary of Radio Terms

Antenna: A metal conductor (often looking like a rod or wire) that helps your radio "speak" (transmit) and "listen" (receive) to radio waves.

APRS (Automatic Packet Reporting System): A digital mode that excels at sharing your location and sending short messages. It's often used for tracking hikers, coordinating during emergencies, or transmitting weather data.

Bandwidth (Wide/Narrow): Think of this as how much space a conversation takes up on the radio "highway." Narrow bandwidth is used in crowded areas to avoid interfering with others.

Call Sign: A unique identifier (like a radio nickname) assigned to licensed amateur radio operators by the FCC. Your call sign must be announced at regular intervals during your transmissions to clearly identify yourself when on the air.

Channel: A pre-programmed memory slot in your radio that stores a specific frequency and other settings. It's like saving your favorite stations in your car radio.

CHIRP: Free computer software that makes programming your Baofeng easier and faster, letting you manage lots of channels and advanced settings.

CTCSS/DCS (Tones): These act like secret passcodes on busy frequencies. By matching the tone of the person or group you want to talk to, your radio will only unmute their transmission, filtering out unwanted chatter.

DMR (Digital Mobile Radio): A popular digital mode used by many ham radio enthusiasts, known for its clear audio and features like text messaging and GPS location data.

Dual Display (A/B): This feature allows you to see two frequencies simultaneously on your Baofeng's display, useful for monitoring an emergency channel while talking on another.

FCC (Federal Communications Commission): The US government agency responsible for regulating the use of radio frequencies. The FCC sets rules for licensing, allowable transmit power, permissible content, and the allocation of different frequency ranges for various purposes.

Frequency: A specific "channel" on the invisible television of radio waves. Different frequencies are used for police, weather broadcasts, ham radio conversations, and more.

Gain (Antenna): A measure of how much stronger an antenna can make your radio's transmitted signal or how much better it can "hear" weak signals compared to a basic antenna.

Manual Programming (On Radio): Using the keypad and screen of your Baofeng to enter frequency information and navigate its menus for storing channels.

Memory Channel: A storage location within your Baofeng that holds a specific frequency and other settings you can recall quickly, like a preset radio station.

Memory Mode (MR): A setting on your Baofeng that uses the pre-programmed channels instead of manually typing in frequencies.

NOAA (National Oceanic and Atmospheric Administration): A US government agency that provides weather forecasts and alerts. Your Baofeng can often receive NOAA weather broadcasts on specific frequencies.

Offset (Repeaters): A special setting for using repeaters, which are like radio relay towers that boost your signal to cover longer distances.

Programming Cable: A special cord that connects your Baofeng to a computer, allowing you to use software like CHIRP for easier channel management.

Public Safety: Refers to designated radio frequencies used by police, fire, and emergency services.

RX: This indicator light or display symbol shows your Baofeng is currently receiving a transmission.

Signal Strength Indicator: Typically bars on your Baofeng's display that show the strength of the signal you're receiving. More bars generally mean clearer audio.

Squelch: A control that filters out background static. It keeps your radio silent until it hears a strong enough signal.

TNC (Terminal Node Controller): A device that acts as a translator between your Baofeng radio and the world of digital data. It converts text, location coordinates, or other information into audio tones that your radio can transmit. On the receiving end, it takes those audio tones back and turns them into readable data for your computer or other devices. Think of it as the bridge enabling your Baofeng to participate in specialized communication protocols like APRS and packet radio.

TX: This indicator light or display symbol shows your Baofeng is currently transmitting a signal.

Transmit Power (High/Low): How much power your radio uses when sending signals. High power gives greater range but drains your battery faster.

VFO Mode (Frequency Mode): A setting on your Baofeng where you type in frequencies directly using the keypad.

Weather Forecasts: Continuously broadcasted weather information on a designated frequency. Your Baofeng can pick these up without relying on internet or phone service.

Chapter 1: A Beginner's Guide to Setup and Basic Functions

Welcome to the World of Baofeng!

You've taken the first step into a fascinating world of communication possibilities by getting your hands on a Baofeng radio. These compact and surprisingly affordable radios offer features you might associate with much pricier gear. But, with all that potential comes a little bit of a learning curve. Let's break things down and have you up and running with the basics in no time.

1.1 The Unboxing Experience: What's in the Box

Opening up your Baofeng package is a bit like finding a toolkit. You'll likely have these main components:

The Baofeng Radio: The handheld unit itself – this is the heart of the system.

Battery: This powers your radio and needs to be charged to start your adventures.

Antenna: That flexible, often rubbery "whip" antenna allows your radio to send and receive signals.

Charger: A cradle or dock where you'll place your battery to replenish its power.

User Manual: Don't toss this aside! It's got detailed info—even if it can be a bit dense to read at first.

Optional Accessories: Depending on your specific Baofeng model and where you bought it, you might have things like an earpiece/microphone, programming cable, or different style antennas.

1.2 Initial Setup: The Essentials

Let's get the most important tasks out of the way to transform your Baofeng from a fancy paperweight into a communication wonder.

Charge That Battery!

Your battery probably has a bit of charge from the factory, but fully topping it off before first use is essential. Proper charging and battery care will significantly impact how much playtime you'll get out of your Baofeng. Locate your charger and carefully slide the battery into it. You're most likely to see a red light turn on, confirming things are happening as they should. When that light turns green, you've got a full charge and the real fun can begin.

Attaching the Antenna

Think of your antenna as helping your radio "speak" and "listen". Find the small metal connector (it looks like a stud) at the top of your Baofeng and carefully align the base of your antenna with it. Give it a clockwise twist – securely tightened but be careful not to overtighten or force it.

Powering On: It's Alive!

Find the power button, which is also typically your volume control knob. Turn it, and your Baofeng should boot up! Its screen will illuminate, and you might even hear a short welcome beep. This is a good sign - you're on your way to mastering the airwaves!

1.3 Navigating the Basics

Let's look at a few important things you'll interact with immediately on your new radio:

The Keypad: It might feel overwhelming at first, but you'll primarily use a few essential buttons right now.

Number Keys: For directly entering frequencies

Menu: Access settings and less-frequently used features

VFO/MR: Switch between frequency (`VFO`) and channel (`MR`) modes (we'll go deeper on these later!)

A/B: Useful for monitoring two frequencies at once

The Display: Decoding the Information

Big Numbers: This is the most prominent piece of info – your current frequency. It is the radio equivalent of a TV channel.

Channel Mode: If you see "CH-" instead of numbers, your Baofeng is in channel mode (using pre-programmed memories).

Icons: Don't panic about deciphering every tiny symbol yet. Important ones for now are likely signal strength and battery life.

Switching Between VFO and Memory Mode

Baofengs give you two main ways to operate. VFO (Frequency Mode) lets you punch in a frequency like typing a TV channel. MR (Memory Mode) uses channels that have been pre-programmed in your radio's memory. Your VFO/MR button toggles between the two.

1.4 Your First Transmission (If Legal)

IMPORTANT DISCLAIMER: Before hitting that transmit button, ensure you understand the licensing laws governing radio use in your location. Transmitting on restricted frequencies without proper authorization can lead to legal trouble!

If you're cleared to transmit, here's the very basic rundown:

1. Frequency Check: Tune to a known frequency used in your area and listen for a bit. Don't interrupt active conversations!

2. Mic Time: Find the "Push-To-Talk" (PTT) button, normally a larger button on the side of your radio. While holding it, speak clearly into the microphone at the top (or your attached earpiece/mic).

3. Release and Listen: Let go of the PTT button. Radio is about taking turns to talk!

Chapter 2: Understanding Buttons, Menus, and Displays

Example: **Baofeng BF-H7**

Product Function

1. Antenna
2. Channel knob
3. Volume / power switch
4. Speaker
5. PTT A
6. PTTB
7. Side button one
8. Side key two
9. Headphone / speaker interface
10. LCD screen
11. Numeric keyboard
12. Battery

Think of your Baofeng's control panel as the cockpit of your communication adventures. While it might seem overwhelming at first, familiarity with the various buttons and understanding what appears on your display will empower you to navigate the airwaves with confidence.

2.1 The Keypad: Your Communication Command Center

Let's break down the most common buttons you'll encounter on your Baofeng and how they are used:

Number Keys (0-9): These are the backbone of your communication.

VFO Mode: Use them to directly enter the frequency you'd like to tune into. It's sort of like manually typing in a TV channel number.

Menu Functions: Numbers sometimes correspond with various menu options to help you quickly select different settings.

Menu:

Accessing this button leads you into a treasure trove of customizations for your Baofeng. Think of it as your radio's hidden settings and control panel. We'll dive deeper into specific menu options later, but some essential functions found here include:

Squelch Control (filtering out background noise)

Power Settings (High/Low output)

Scan Settings (Searching for active frequencies)

VFO/MR: Remember this crucial button from Chapter 1? This toggles between:

VFO (Frequency Mode): This is where you directly punch in frequencies using the numerical keypad.

MR (Memory/Channel Mode): Here, your radio uses pre-programmed channels that store frequency information and other settings. Consider them communication bookmarks.

A/B: Enables you to simultaneously monitor two frequencies. Picture it as listening to two channels at once! This is especially useful if you want to keep an ear on an emergency channel while talking with a friend on another frequency.

CALL: The transmission button. Holding it down allows you to speak into the microphone. Remember, always listen first to ensure the frequency is clear before transmitting!

Other Function Keys: Baofeng models offer other keys for activating a flashlight, toggling through specific settings, or quick access to commonly used functions. Refer to your manual to learn about any special keys on your device.

2.2 Your Display: A Window into the Radio World

The screen of your Baofeng, while compact, serves as your central dashboard, providing critical information. Here's how to understand what it displays:

Main Frequency/Channel: This large number (or channel name) represents the frequency or channel your radio is currently tuned to.

Dual Display (A/B): Displays two frequencies when you're actively monitoring two.

Signal Strength Indicator: Typically a series of bars that increase or decrease. Consider this your "voice quality" meter – more bars generally mean clearer audio.

Battery Indicator: Your little fuel gauge for the radio, often visualized as a battery icon. It usually features several bars that deplete as your battery drains.

TX/RX: "TX" appears when you're transmitting, "RX" when receiving a signal.

Other Icons: A multitude of tiny symbols can display. These signify features like scan mode, keypad lock, etc. Your manual will provide a key for identifying and understanding these.

2.3 Navigating the Menus: Exploring the Possibilities

The menu system unlocks countless features to customize your Baofeng. While there might be subtle differences across models, some common menu options include:

SQL (Squelch): Allows you to fine-tune noise filtering, preventing static bursts when there's no active signal.

STEP (Frequency Increment): Defines the "jump" amount when changing frequencies (e.g., 5 kHz, 25 kHz). This impacts how quickly you move across the radio spectrum.

TXP (Transmit Power): Enables you to toggle between Low/High power modes. Lower power might conserve battery life while high power could give you greater range under certain conditions.

(And Many More!) Features controlling backlight, keypad tones, voice prompts, and other advanced settings reside within the menu system.

Important: Always consult your manual for precise menu layout and functions specific to your Baofeng model.

2.4 Practical Menu Example: Fine-Tuning Your Squelch

One of the most frustrating things when using any radio is hearing a burst of static every time a weak signal fades out. Your Baofeng's "squelch" function saves you from this! Fine-tuned squelch cuts out unnecessary background noise, improving clarity.

Here's a typical, simplified approach to adjusting your squelch using the Baofeng menu system:

Enter the Menu: Press the "Menu" button on your Baofeng.

Find the Squelch Setting: Different Baofeng models use varying numbers to represent specific menu items. Refer to your manual to find the menu number leading to "SQL" or "Squelch." For example, this setting might be accessible by pressing "Menu," then "2," then "0."

Adjusting the Squelch Level: The squelch is adjustable on a scale – with "0" usually allowing all signals through (even the weakest, noisy ones) and "9" being the strictest setting (cutting out most background noise). Experimentation is key here. Typically, "5" is a good starting point. After setting a chosen level, press "Menu" to go back a level.

Exit the Menu: Press the "Exit" button (sometimes identified by an arrow) repeatedly until you return to the main display screen.

Test it Out: Tune to a frequency where you usually receive signals, but there's no one talking at the moment. If the squelch works well, you'll have blessed silence. When someone begins transmitting, their voice should easily overpower the squelch and come through clearly.

Navigating Your Way to Success:

Let's imagine you want to increase squelch slightly because you find there's still some occasional static crackling through. Here's a potential workflow:

Press "Menu," then enter the numbers that take you to the squelch setting. Let's assume it's "Menu → 2 → 0." Your screen should display something like "SQL – 5".

Use the keypad to set a slightly higher squelch, like "6."

Press "Menu," then "Exit" to return to the main screen.

Listen in; If the background noise has disappeared, success! If not, you might repeat this process, increasing the squelch another notch.

Notes:

Each Baofeng model navigates menus slightly differently. Understanding your specific manual is truly the key to mastery.

Squelch is a balance: Too much silence and a strong, but distant, signal might remain unheard! Too little, and your listening suffers.

Practice Makes Perfect

Exploring different menu items is the best way to become familiar with your Baofeng. Even if you don't want to immediately change anything, scrolling through your menu options shows you what's possible!

Chapter 3: Frequencies, Channels, and CHIRP Basics

While you can manually punch in frequencies on your Baofeng to start tuning in, true flexibility and ease of use come from effectively programming and storing essential frequency information within your radio. This chapter covers that and introduces the powerful world of CHIRP software.

3.1 Understanding Frequencies: The Foundation of Communication

Think of radio frequencies like "channels" on a giant invisible TV set. Specific frequencies are assigned for various purposes:

Public Safety: Police, fire, emergency services operate on dedicated frequencies. It's vital to listen but to never transmit on these unless a genuine emergency demands it.

Amateur Radio (Ham): Requires licensing but grants access to an enormous worldwide network. Hams use special frequencies for hobby communication, emergency preparedness, and experimentation.

Weather Forecasts: Broadcast continuously on a designated frequency. Your Baofeng can provide up-to-date alerts without relying on phone or internet connectivity.

Marine: Boaters communicate with each other and authorities on separate channels to ensure safety and coordinate with coastguard and harbor patrols.

And more! Numerous frequencies offer businesses, private groups, and specialized applications.

3.2 Memory Channels: Communication Presets

Manually entering frequencies every time you want to switch between different 'conversations' or communication services would be tedious. Luckily, Baofengs have "memory channels." Essentially, memory channels let you program a frequency along with various settings and store them for one-button access.

Think of this as presetting stations on your car radio for quick tuning changes!

3.3 Programming Methods: Manual vs. Software

There are two primary ways to get those coveted frequencies and settings into your Baofeng:

Manual Programming (On Radio):

It involves using the keypad and screen to enter frequency information and navigate its menus.

Pros: You can do this anywhere – no computer needed.

Cons: Time-consuming, especially for many channels.

CHIRP Software (Computer):

Free, open-source software that connects your Baofeng to your computer via a special (usually purchased separately) programming cable.

Pros: Lightning-fast programming of numerous channels, easier editing, ability to download pre-made channel lists.

Cons: Requires initial setup, cable, and a basic understanding of the software.

3.4 Mastering CHIRP: Your Programming Powerhouse

While both programming methods are valid, CHIRP offers far greater speed and customization for serious Baofeng users. Let's outline the basic process:

Downloading and Installing: Start by downloading CHIRP from https://chirp.danplanet.com/projects/chirp/wiki/Download and installing it on your computer.

Connecting Your Baofeng: Utilize the special programming cable, connect one end to your Baofeng radio (usually a headphone-like jack) and the other end to a USB port on your computer.

Download from Radio: After launching CHIRP, use the built-in functions to "Download from Radio." This creates a file containing your current Baofeng's programming, acting as a backup or starting point.

Adding Channels: CHIRP gives two paths:

Manual Entry: Type in frequency details alongside additional settings like transmit power, names, and other specific details.

Import: CHIRP can load ready-to-use lists of channels created by others or found online, greatly accelerating setup (ensure your region uses matching frequencies!).

Upload to Radio: Once channels are added or modified, choose the "Upload to Radio" option in CHIRP, transferring your curated configuration to your Baofeng.

Important Resources

CHIRP Website: (https://chirp.danplanet.com) – Offers download links, instructions specific to your radio model, and a support forum.

Your Baofeng Manual: Guides you through connecting the programming cable and any special settings needed for computer connection.

3.5 CHIRP Step-by-Step: Programming Made Easy

Let's assume you've installed CHIRP and have your Baofeng connected. A simplified initial session to add a few channels might look like this:

Create New File: Open CHIRP and select "File -> New." This prompts you to choose your Baofeng's exact model. Select it from the offered choices, creating a blank workspace for your radio's configurations.

Channels by Hand: Navigate to the "Memories" tab. Right-click and select "New Channel." A window pops up. Fill in the "Frequency," give it a memorable "Name," and choose options like "Transmit Power" (usually start with "Low"). Press "OK" to save, then repeat for a few more essential frequencies.

Upload: Click the arrow icon pointing towards a radio at the top of the CHIRP window. Confirm you want to "Upload to Radio." If everything is connected and set properly, all your new channels are now swiftly programmed into your Baofeng!

3.6 Understanding Channel Settings for Optimal Communication

Beyond basic frequency and name, each channel stored in your Baofeng has an array of customizable settings. Exploring these allows you to fine-tune your experience. Here's a breakdown of some key ones:

TX Power (Transmit Power): Impacts transmission distance. "High" allows longer reach but drains your battery faster. "Low" is suitable for local conversations and extends talk time.

Wide/Narrow Bandwidth: Changes sound quality and how wide a section of the radio spectrum is used. Select "Narrow" for more crowded or congested scenarios to avoid interfering with others.

CTCSS / DCS (Tones): These filter out unwanted conversations on shared frequencies. If you want to only hear from a specific group or radio, you'll need to match their unique tone settings.

Offsets (For Repeaters): Specialized settings for using repeaters – devices that boost signals, extending your range. This will be further explored in advanced chapters related to specific applications.

Chapter 4: Choosing the Right Tool for the Job

"Why does my radio have such a terrible range?" An underperforming antenna is often the culprit, and the right antenna can significantly boost your Baofeng's ability to both send and receive signals. This chapter demystifies antennas and enables you to select the perfect antenna sidekick to amplify your communication possibilities.

4.1 Antenna Fundamentals: It's Not Just a "Whip"

At its core, an antenna is a conductor designed to radiate (transmit) and capture (receive) radio waves. Let's break down some key things to understand:

Size Matters (To an Extent): Wavelength - the physical distance a radio wave travels during one cycle - is relevant here. Matching antenna length to wavelength often increases efficiency (although Baofeng antennas utilize clever 'hacks' and loading techniques to avoid needing skyscraper-sized rods due to the frequencies they use!).

It's About Energy Transmission: Think of your antenna as a bridge between your Baofeng's electrical signals and the invisible world of radio waves. Effective antennas convert those signals into waves with minimal loss and also act to catch these waves and efficiently relay them to your radio.

Not One Size Fits All: There's no single antenna perfect for every scenario. Terrain, desired range, and specific frequency ranges all influence the ideal choice.

4.2 Deciphering the Stock Antenna

The little rubber "ducky" antenna included with your Baofeng is surprisingly robust but lacks the power and performance potential of specialized alternatives.

Pros: Flexible, portable, great for very general, close-range use.

Cons: Inefficient on certain frequencies, easily obstructed (body, walls, etc.), and limited range.

It's a Starting Point: Consider your stock antenna as your baseline. As you explore other options, you'll be stunned by how much improvement is achievable.

4.3 Types of Antennas for your Baofeng

Let's explore common antenna categories to enhance your Baofeng experience:

Upgraded Whip Antennas: Longer, thicker variations on the 'rubber ducky' theme, improving performance. (Some offer telescoping designs for portability AND performance boosts)

Magnetic Mount (Mobile): For vehicles. The powerful base magnet ensures stability, boosting both transmit and receive capability.

Directional Antennas (Yagi): Beam focused radio energy in a specific direction. Allows reaching distant contacts or precise targeting when lots of radio "noise" exists.

Base Station Antennas: Mounted high and often larger, these provide home bases with a wider reach. Think of them as radio lighthouses for reliable long-distance contact.

4.4 Factors to Consider for the Right Fit

Frequencies in Use: Antennas optimized for certain frequency bands exist. Ensure compatibility!

Gain: Think of this as an antenna's power multiplier. Measured in "dB," more gain generally means increased range but possibly a more focused direction.

Environment: Is this for on-the-go hiking use, mounted on a vehicle, or part of a stationary setup? This helps guide your size and mounting needs.

4.5 Beyond the Basics: DIY & Modification, Optimization Tools, and Further Learning

The thrill of optimizing your Baofeng experience extends beyond readily available antennas. Embracing the DIY spirit with antenna projects saves you money, allows for hyper-tailored results to match your exact situation, and offers immense satisfaction through self-reliance. Projects might seem daunting, but it's a journey into exciting territory! Simple wire antennas like the J-Pole (available on numerous resources such as https://www.instructables.com) demonstrate how easily sourced materials yield astonishing results. The creative potential stretches from repurposing tape measures and metal hangers to transforming old satellite dishes with a little effort and research on specialized enthusiast forums. It's an exploration where innovation knows few bounds!

However, building is just the start. To unlock the true potential of both your own antennas and prebuilt options, analysis and optimization tools are key. Antenna analyzers provide a window into an antenna's resonant frequency, ensuring efficiency and performance. These range from affordable DIY versions to comprehensive professional units tailored to precise, experimental purposes. Beyond hardware, digital modeling software such as MMANA-GAL lets you virtually test complex designs and setups before the build phase, saving precious time and resources. Additionally, SWR meters become an essential partner by measuring "Standing Wave Ratio." These insights allow you to perfect the match between your Baofeng and antenna, minimizing loss and boosting performance – essential for both DIY and commercial antennas.

The pursuit of antenna knowledge thrives within a strong community. Whether seeking local amateur radio clubs or venturing into online forums and repositories like The ARRL Handbook, you'll encounter fellow enthusiasts willing to share secrets, guide your building pursuits, and help fine-tune your setups. These valuable sources of information serve as launchpads for expanding your antenna arsenal and mastering the principles behind this intricate, impactful part of your Baofeng journey.

Example 1: The J-Pole Antenna

Why It's Popular: The J-Pole offers amazing performance-to-cost ratio, easy construction, and solid effectiveness across certain VHF and UHF frequency bands.

The Design: Typically constructed from copper pipe or twin-lead TV antenna wire, it uses a clever configuration to achieve resonance without a vast structural footprint.

Resources:

Step-by-Step Plan on Instructables ([invalid URL removed]): Great beginner-friendly with photos and clear material suggestions.

"KJ4IZW on Antennas" Youtube Channel: ([invalid URL removed]) Deep dives into building various J-Pole variants.

Example 2: The Tape Measure Yagi

Why It's Popular: Shockingly effective yet requires minimal tools. Great for portable adventures where high-gain directional performance is needed.

The Design: Relies on cleverly measured lengths of conductive tape adhered to a non-conductive central member (wooden bar, PVC pipe, etc.) It achieves signal directivity, boosting a signal toward a focused area.

Resources:

DK3RED Website Yagi Tape Measure Project Guide: ([invalid URL removed]) Offers precise calculations, a focus on UHF, but principles are adaptable.

Youtube Search: Search "Tape Measure Yagi Baofeng" to uncover variations on materials and a visual approach to construction.

Example 3: Repurposed Coaxial Cable Antenna

Why It's Popular: Creative use of discarded or easily acquired resources - great for experimenting without high investment.

The Design: Specific design varies wildly! The principle leverages the coax cable's shielding and core. Precise cutting and configuration create various radiation patterns.

Resources:

Internet Searches are Paramount: This is more advanced - search phrases like "DIY Coax Dipole Baofeng" or "Baofeng repurposed coax antenna design". Be prepared to filter numerous results and adapt what you find!

Building Notes

Frequency Matters: Each design performs ideally on specific bands. Always factor in your primary usage!

Calculators are Your Friends: Various online calculators exist (search "J-Pole calculator," etc.) that ease calculations, saving you from complex formulas.

Start Simple: Choose projects with abundant guides, matching your initial skill level. Success breeds experimentation!

Seek Community: Forums and local radio clubs welcome newbie builders - a fantastic support system!

Chapter 5: Tips and Tricks for Clear Communication

Frustration sets in when your Baofeng's transmissions sound weak, garbled, or fail to reach their intended target. While a powerful antenna plays a crucial role, many additional factors influence your signal strength and overall communication quality. In this chapter, we'll uncover simple adjustments, environmental considerations, and a few tricks to transform you from a "barely audible" radio operator to the one with a booming, crystal-clear voice on the airwaves.

The Importance of Environment

Before diving into your Baofeng's settings, let's explore how your surroundings impact radio transmission:

Terrain: Hills, valleys, and dense forests create obstacles for radio waves. Cityscapes with tall buildings also hinder easy signal propagation. Seek high ground whenever possible or consider how your local geography impacts your communication range.

Structures and Interference: While your Baofeng may work well outdoors, performance can suffer significantly within buildings. Metal, concrete, and certain types of insulation are notorious for blocking signals. Furthermore, electrical appliances (power lines, microwave ovens, etc.) can create a noisy radio environment that drowns out weak transmissions. Experiment with operating your radio in different locations within your home or workplace to find the most favorable spots.

Optimize Your Baofeng Setup

Let's focus on squeezing the most performance from your radio itself:

Battery Power: A dying battery translates to a weak signal. Always ensure your Baofeng is sufficiently charged for clarity and maximum output power.

Transmit Power (High vs. Low): Your Baofeng typically offers selectable power settings. Utilize "High" mode for maximum reach when permissible, but switch to "Low" power for close-range local communications. This also extends your battery life.

Squelch Settings: Remember this from earlier chapters? A too-strict squelch can "cut out" weak, yet still understandable, incoming signals. Adjust your squelch to the threshold where background noise is silenced while not impeding reception.

Clarity vs. Range (Narrow vs. Wide): On crowded frequencies, "Narrowband" mode focuses your transmission, minimizing interference. However, "Wideband" can slightly increase your range at the expense of minor audio quality degradation in some situations. Experiment to find the ideal balance.

Body as Barrier

A simple, often overlooked issue - your own body can absorb and block your Baofeng's signal! Avoid holding the radio close to your torso or in a position where you're between the antenna and your desired target. Even small adjustments in how you hold your radio can make a surprising impact!

Advanced Techniques (With Caution)

While we won't explore them in depth here, some enhancements exist for the adventurous:

External Speakers: Improve audio clarity in noisy environments, enhancing your ability to pick out faint signals.

Repeater Networks: These stations rebroadcast your signal. When strategically positioned on high ground, they dramatically extend your Baofeng's range. However, their use involves understanding proper licensing and etiquette.

Troubleshooting Checklist

Before frustration sets in, follow these steps if your signal is unexpectedly poor:

Location, Location, Location: Try moving to a slightly higher position or a less obstructed area.

Battery Check: Is it fully charged? Does switching batteries improve the situation?

Antenna Woes: Ensure it's securely attached. If possible, temporarily swap to a known-good antenna to pinpoint problems.

Power & Squelch: Double-check you're transmitting with the correct power settings and that your squelch isn't too strict.

Continuous Improvement

Optimizing your signal is an ongoing process! Keeping a small notebook to log your experiences – noting when communication was successful and when it struggled, alongside details like location, weather, etc. – helps you decode patterns. Over time, you'll transform into an expert on maximizing your Baofeng's effectiveness in any scenario.

The Mobile Advantage... and its Challenges

Operating a Baofeng from your car, truck, or other vehicles offers several advantages:

Height Advantage: Even a car modestly elevates your antenna above ground-level obstructions, potentially boosting range.

Power Source: Easily tap into your vehicle's electrical system for long-lasting operation and consistent maximum power output from your Baofeng.

Enhanced Antenna Options: Vehicle mounts allow significantly larger and more powerful antennas (like magnetic mount whips) compared to handheld situations.

However, vehicles introduce additional considerations for optimal signal:

Metal as a Cage: Your car's bodywork can act like a shield, hindering signals. External antenna mounts circumvent this issue.

Electrical Noise: Vehicles are full of electrical systems with the potential to cause interference. Sometimes, adjusting power sources or strategically grounding parts of your setup mitigates this noise.

Staying Safe and Legal: Ensure your Baofeng setup doesn't impair safe vehicle operation! Laws on radio use while driving might vary in your region, so become familiar with the local regulations.

Mobile Setup Tips

Let's maximize your mobile Baofeng experience:

Invest in a Magnetic Mount: These provide a secure, temporary base for powerful external antennas that dramatically enhance performance.

Placement Matters: Ideally, position your external antenna towards the center of your vehicle's roof for optimal 360-degree radiation patterns.

Grounding Considerations: Online guides for your specific vehicle often recommend the best "ground points" to minimize noise caused by the engine and other electrical components.

Keep it Neat: Invest in proper cable routing to avoid messy or tangled wires that create safety hazards while driving.

A Note on HTs Inside Vehicles

Even without an external antenna, your handheld Baofeng can often outperform a cell phone in rural areas due to the superior power and frequencies it utilizes. However, don't underestimate how much the vehicle blocks its signal! If possible, holding your Baofeng near a window or extending the antenna outside will often improve reception noticeably.

Typical Baofeng Car Setup

Baofeng Transceiver: Securely situated within easy reach of the driver, typically using a dashboard mount or a dedicated holder for convenience and safety.

Power Connection: The Baofeng's power cable is neatly routed and connected to a suitable power source within the vehicle. This might be the cigarette lighter adaptor, directly to the fuse box, or a dedicated power distribution panel, depending on the desired setup complexity.

Coaxial Cable: A high-quality coaxial cable runs from the back of the Baofeng radio, carefully routed to minimize sharp bends and avoid potential pinch points near car doors or moving seat mechanisms. The routing should culminate towards the chosen antenna mounting location.

Magnetic Mount Antenna: Placed as close as possible to the center of the vehicle's metal roof, ensuring a strong magnetic grip for stability.

Antenna and Cable Connection: The end of the coaxial cable connects securely to the base of the magnetic mount antenna. Excess cable is neatly coiled to prevent snags and potential damage.

Where to Find Clear Visuals

Image Search: Utilize search engines with a query like "Baofeng magnetic mount car setup" or "Baofeng mobile radio installation diagram." Filter your results to the "images" tab to quickly uncover numerous visuals.

Retailer Websites: Online stores selling Baofeng gear and magnetic mount antennas will often include product photos or even instructional guides demonstrating a typical installation setup.

Youtube Videos: Search for "Baofeng mobile setup" or "Baofeng car installation" to find video walkthroughs and user demonstrations showcasing the entire process visually.

Important Notes:

Your Specific Vehicle: The best location for the antenna, ideal cable routing, and noise-reducing grounding points all vary between car models. Seek resources for your specific vehicle make and year!

Safety and Legality: Always prioritize secure installation and abide by the laws governing radio use while operating vehicles in your region.

Chapter 6: Effective Communication Techniques

Beyond technical know-how, becoming a respected and sought-after voice on the airwaves requires understanding the established norms and practices of radio communication. Whether you're a casual user, emergency preparedness enthusiast, or seasoned ham operator, mastering etiquette not only makes communication smoother but also contributes to a positive and welcoming experience for both yourself and others sharing the airwaves.

The Golden Rule: Listen First!

Always begin by listening to an active frequency for several minutes before transmitting. This ensures you won't accidentally interrupt an ongoing conversation or emergency transmission. Think of it like politely knocking before entering a room. Even if the channel appears initially silent, take a moment - there might be weak signals on the edge of your radio's reach that need time to emerge.

Clear Identification

Legality often mandates precise identification when transmitting. This includes:

Ham Radio Operators: State your assigned call sign at regular intervals (as dictated by regulations) and at the start and end of your communication.

Others (GMRS, etc.): While specific rules vary, identifying yourself with a name or chosen identifier allows others to address you directly. It also fosters accountability and civility

Clarity and Conciseness

Radio communication isn't a casual phone chat. Speak clearly, enunciating your words for maximum clarity, especially amidst noisy environments. Keep transmissions brief and to the point. Think about what you want to say before pressing the transmit button.

Avoid Jargon (Initially): While groups have their own lingo, start with plain language for newcomers to feel included.

Using Your Mic Right

Distance Matters: Hold it a few inches from your mouth, avoiding overly loud or quiet transmissions.

Avoid Background Noise: Find a quiet spot, away from wind gusts, machinery, or side conversations disrupting others.

Think Before Keying Up: Avoid the urge to "hold" your channel by pressing the transmit button without speaking – respect others ready to use the frequency.

Common courtesy in Practice

Welcoming New Voices: If you hear someone struggling or unfamiliar with the usual flow, offer assistance kindly. We all started somewhere!

Leaving Space: Provide short breaks between transmissions, allowing others to break in if it's time-sensitive.

Emergency Priority: If you hear "emergency traffic" announced, immediately cease non-urgent transmissions and monitor in case assistance is needed.

Respect the Channel: Certain frequencies have established purposes (weather updates, formal nets, etc.). Know these before transmitting.

Pro-Words for Efficiency

Specific phrases are common on radio to streamline communication:

"Break-Break": Indicates urgent traffic or an immediate need to interrupt.

"Over": Signals you've finished your part of the exchange and await a reply.

"Roger": Means "message received and understood" (not simply "yes").

Phonetic Alphabet: For precise spelling of call signs, locations, etc., especially in poor signal conditions. Know the standard (Alpha, Bravo, etc.)

Resources – Going Beyond the Basics

Local Clubs: Seek out radio clubs in your area. They're repositories of best practices and welcoming to teaching good habits.

Formal Guides: The ARRL (or your country's equivalent) often publishes guides on etiquette and procedures for specialized communication types.

Monitoring Is Learning: Spend time simply listening to various frequencies to observe established communication styles.

Continuous Improvement

Like any skill, radio etiquette takes practice. Don't be discouraged by the occasional misstep! Over time you'll embody these principles, earning respect on the airwaves and finding interactions smoother and more fulfilling.

Example: You've been monitoring a frequency where a group of individuals or stations are having an organized conversation (perhaps a local club conducting a weekly meeting or an emergency response training net).

Steps for Respectful Entry:

Continue Listening: Give yourself more time to grasp the flow of the conversation – who seems to be acting as the "net controller" (leader), the topic at hand, and the pace of call sign exchanges.

Identify an Opening: Wait for a clear pause between transmissions during which a previous speaker has likely concluded with an "over" or similar phrase indicating their turn is done.

Brief Transmission: Key your mic and clearly announce "This is [Your call sign/identifier], standing by." Keep it concise!

Awaiting Acknowledgement: The net controller or another member may acknowledge you directly ("[Your Call Sign], go ahead") or may indicate that the net is currently closed to new traffic. Be patient.

Follow Instructions: If allowed to join, follow any directions given for when you can formally introduce yourself or participate fully. If not, politely acknowledge ("[Your Call Sign] understood, monitoring only") and continue listening.

Key Points:

Avoid Barging In: Never interrupt an in-progress transmission. A seemingly quiet channel might simply be a temporary pause between speakers.

"Standing By" is Polite: It shows your respect for the ongoing activity and willingness to wait your turn.

Brevity is Key: Lengthy introductions during your initial transmission disrupt the flow of the established net.

Additional Tips

If Unsure, Keep Monitoring: It's perfectly acceptable to remain in a listening mode until you're more confident about joining. Often, you'll glean valuable information by just observing.

Formal Nets: If the net is highly structured (specific check-in procedures, etc.), seek out information from the operating club or online guides prior to your first attempt at participation.

Before Check-In:

Monitoring for [Net Name]: Indicates you're listening to the frequency and awaiting the proper time to check in.

Ready for check-in: Lets the net control know you're prepared to begin the check-in process.

During Check-In

[Net Name], this is [Your Callsign/Identifier]: Formal introduction with your assigned callsign or chosen identifier.

[Location]: States your geographic position (city, general region, etc.).

[Signal Report]: Provides an assessment of your signal strength and clarity on their end (common format is "5 by 5" for excellent).

[Additional Info]: If relevant, you might include details such as radio type, antenna in use, and operating mode.

Example Full Check-In

"Monitoring for Alpha Net. Ready for check-in."

"Alpha Net, this is W1XYZ. Boston, 5 by 5."

Ending Check-In

73: Standard sign-off phrase used at the conclusion of a communication.

Thank you for the check-in: Acknowledgment by the net control.

Courtesy Phrases

Standing by for instructions: Shows your readiness to receive directions from net control.

Any traffic for me? Inquiring about messages or communications specifically meant for you.

Thank you for the call: Expresses gratitude for the opportunity to participate.

Additional Useful Terms

Net Control: The person managing the net's operation.

Over: Signifies that you're finished speaking and are awaiting a response.

Break: Indicates the need to interrupt the communication due to an urgent matter.

Roger: Confirms that you have received and understood a message.

Tips:

Observe carefully and follow the instructions given by net control.

Be concise and clear in your transmissions.

Utilize a polite and friendly tone of voice.

Using these phrases and familiarizing yourself with check-in procedures will help ensure efficient and smooth communication during your participation in radio nets.

Chapter 7: Using Your Baofeng in Critical Situations

While Baofeng radios excel in daily adventures, their true value shines in the face of crisis. From localized outages to widespread disasters, independent communication becomes a lifeline. Being prepared isn't just about owning a radio– it's a mindset. Let's transform your Baofeng into a vital preparedness tool.

Start by knowing what frequencies matter ahead of time. Research channels dedicated to emergency services, weather reports, and any local volunteer networks within your area. Programming these into your Baofeng in advance saves precious time when seconds count. If you're part of a family, neighborhood, or preparedness group, a communication plan is crucial. Decide on backup meeting places, the frequencies you'll monitor, and pre-arranged times to check in. It's one thing to have powerful equipment, but it's another to know how and when to use it effectively. Practice regularly to ensure everyone involved is comfortable with basic Baofeng operation – a crisis is no time for a user manual crash course!

Power outages often coincide with emergencies. Keep multiple batteries charged and consider solar or hand-crank generators as backups for extended situations. Your car can serve as a mobile base station; adapters allow you to tap into its battery, run your Baofeng, and even boost your signal range with a larger antenna. However, conserve power whenever possible! Use "Low" power mode, disable unnecessary features like backlights, and keep transmissions concise to make the most of your battery life.

During an emergency, your Baofeng becomes your ears on the world. Utilize it to monitor emergency channels for official updates and instructions, as well as any communication networks established by local aid groups or search and rescue teams. Up-to-date weather reports are vital – tune in to dedicated NOAA weather channels for advance warnings of dangerous conditions.

When staying connected could make all the difference, keep these communication principles in mind: clear call signs (even improvised ones if unlicensed in a true emergency) and concise speaking are paramount for clarity. Consider learning about repeater networks that extend your range but be aware of their access procedures and proper etiquette. If allowable by your license and situation, explore text-based modes (such as APRS or DMR) that can cut through the noise when voice transmission becomes difficult.

Understanding your Baofeng's limitations is also part of responsible preparedness. Before an emergency, familiarize yourself with regulations regarding radio use in your region – what's normally restricted may be permitted in crisis scenarios. In chaos, radio channels become crowded; respect ongoing emergency traffic and never interrupt unless you have critical information. Furthermore, verify what you hear against reliable sources before acting – misinformation runs rampant in uncertain times.

Seeking out further resources bolsters your preparedness. Connect with local preparedness groups for practices specific to your area, tap into the knowledge base of the ARRL (or your country's equivalent), and investigate resources provided by governmental agencies like FEMA (or its counterpart). These organizations often provide specialized guides and training to hone your crisis communication skills.

Scenario 1: Severe Thunderstorm

The sky darkens, the first rumble of thunder sounds... and then the downpour begins. Severe thunderstorms pack more than just rain - power outages, flash flooding, and wind damage pose significant risks. Here's where your Baofeng comes into play:

Weather on Demand: Don't solely rely on your phone! Tune into NOAA weather channels pre-programmed into your Baofeng for real-time alerts, radar updates, and specific warnings tailored to your exact location. Knowledge is preparedness!

Neighbourhood Network: If you've established a community communication plan, touch base with designated neighbours on your agreed-upon frequency. Exchange information about localized flooding, fallen trees, or anyone in need of immediate assistance.

Reporting Hazards: Spot a downed power line sparking? A tree about to block a main road? Contact your local non-emergency line or, if pre-established, the relevant volunteer response team via your Baofeng to report the situation. Clear communication may prevent further damage or accidents.

Baofeng Tips for the Storm

Indoor Antenna Issues: If you must stay inside, but your signal degrades, see if running the antenna cable towards a window improves reception slightly.

Helping those Less Prepared: Elderly neighbours may not have up-to-date weather information. If safe to do so, offer to relay critical warnings or check if they need supplies. Your Baofeng aids in this good deed!

Power Loss Prep: Is a prolonged outage likely? Switch your Baofeng to "Low" power mode and consider relying on a pre-charged battery to preserve your main one for when the storm subsides.

Scenario 2: Power Outage During a Heatwave

Losing power is always frustrating, but in extreme heat, it raises health concerns. Let's examine how your Baofeng becomes an essential lifeline:

Official Update Hunt: Search for frequencies used by your local city or power company. They may transmit estimates for restoration times, cooling shelter locations, or water distribution points.

Community Connection: Neighbourhood networks become crucial! Coordinate with others regarding safe water sources, sharing portable generators (if compatible with your Baofeng), and checking on anyone particularly vulnerable to the heat.

Conserving Baofeng Power: Use "Low" mode, short transmissions, and disable extra features. Consider a hand-crank or solar charger for extended outages. Rotating batteries keeps your communications flowing, especially for vital check-ins.

Beyond the Basics

Vulnerable Neighbours: Is someone nearby reliant on a medical device requiring power? If it's safe, relay their situation to emergency channels on your Baofeng or to neighbours who may be able to render aid.

Heat Safety Information: Share reliable sources for heat safety tips over a local frequency. Your Baofeng can be a tool to combat the risk of heat exhaustion and heatstroke for your community.

Important Note: Always prioritize your own safety and that of others. Don't venture out in dangerous weather solely to provide radio updates unless you're part of an organized, trained response team.

Scenario 3: Targeted Incident

A news alert flashes across your phone: an active event at a nearby school, public gathering, or a major workplace. Panic is a natural response, but your Baofeng empowers you to go beyond the initial fear and take informed action.

Monitoring Official Channels: Tune to designated emergency service and local law enforcement frequencies. Avoid clogging these channels with transmissions, but actively listen for reliable information, potential evacuation orders, and developing updates on the situation.

Connecting with Loved Ones: If cell networks are overwhelmed, your Baofeng might offer a way to reach family members near the incident. Pre-arranged frequencies and call signs streamline communication amidst chaos. Focus on short transmissions of well-being and location updates to conserve your battery and the airwaves.

Combating Misinformation: In tense situations, rumors fly. Before relaying anything you hear over your Baofeng, make every effort to verify the information's source. Prioritize messages coming from official channels, and if in doubt, refrain from further spreading unconfirmed details that could cause additional panic or misdirected resources.

Baofeng Considerations During an Incident

Discretion and Sensitivity: Be aware of your surroundings and avoid transmitting near active emergency personnel if possible. Their communication takes priority while ensuring your safety.

Group Coordination: If you're part of a neighborhood watch or workplace preparedness plan, your Baofeng connects you with the wider network. Follow any established communication protocol for reporting observations, assisting those in need discreetly, and relaying information to a central point if applicable.

Mental Preparedness: Localized incidents are deeply stressful. Remember, you're not alone! Community networks or even just monitoring emergency channels for updates can provide a sense of connection vital in maintaining composure.

Scenario 4: Hurricane/Wildfire Threat

The advance warning of a major hurricane or the encroaching wildfire front presents a unique set of challenges. Let's see how your Baofeng becomes your link to preparation and potential evacuation coordination.

Weather Updates Are Key: Tune in to NOAA weather alerts and fire monitoring frequencies depending on the threat. These provide targeted information to your location and dictate the urgency of your preparations.

Preparedness Group Power: Actively monitor any community groups or volunteer organizations on established frequencies. Information about supplies, evacuation staging areas, or changes to routes can be relayed quickly via radio when traditional communication channels may fail as the threat draws closer.

Assisting Evacuation: If ordered to evacuate, use your Baofeng to coordinate with neighbors and preparedness groups. Sharing carpooling options, identifying those needing mobility assistance, and maintaining group communication during the actual evacuation can make a chaotic situation smoother.

Additional Considerations for Large Disasters

Power Plan: Prolonged evacuation may mean seeking shelter outside of normal power grids. Invest in solar charging options or consider acquiring car adapters to maintain your Baofeng's usefulness.

Beyond Voice: If licensed, explore text-based digital modes. These can transmit further or cut through in congested conditions when voice communication gets difficult.

The Importance of Training

In both of these high-stress scenarios, the value of pre-established protocols and practice with your Baofeng skyrockets. Knowing frequencies, communication plans, and the radio's operation instinctively frees your mind to focus on the larger situation.

Scenario 5: Social Unrest

Heightened tensions within a community can erupt into civil disturbances, potentially disrupting phone and internet services. Your Baofeng offers alternative ways to stay informed and ensure safety.

Situational Awareness: Monitor designated emergency channels and frequencies used by local law enforcement. This provides a real-time picture of changing conditions, potential areas to avoid, and any developing curfews or safety restrictions.

Staying Safe: If part of a neighborhood watch or community network, use your Baofeng to coordinate check-ins, share verified observations, and report any escalating situations that require immediate response. Strength lies in organized, informed communication during times of instability.

Connecting with Aid Networks: If the situation grows more severe, pre-established channels used by humanitarian aid groups or volunteer organizations become invaluable. Your Baofeng may be your way of requesting assistance, reporting injuries, or finding vital resources if traditional support systems become overwhelmed.

Responsible Baofeng Use Amidst Unrest

Prioritize Safety: Never place yourself in harm's way solely to provide radio updates. Monitor actively, but use discretion in when and what you transmit, especially when sensitive information is involved.

Verify Before Relaying: Misinformation spreads rampantly during unrest. Double-check what you hear over the radio against other reliable sources if possible. Amplifying rumors fuels confusion and can lead to further destabilization.

Respectful Communication: Even in disagreement, maintain civility on the airwaves. Your Baofeng isn't a platform to escalate tensions but rather a tool to build bridges when possible and focus on the safety of all involved.

Scenario 6: Supply Chain Issues

When access to food, fuel, or medicine suffers, it goes beyond mere inconvenience; it's a community survival issue. Let's see how your Baofeng can contribute towards finding solutions and mutual support.

Resource Sharing Networks: Coordinate with neighbors or pre-established community groups to set up information sharing frequencies. Reports of where dwindling resources are found, potential supply runs to outlying areas, or even identifying those with critical medical needs become easier with organized radio communication.

Bartering and Mutual Aid: Your Baofeng has the potential to facilitate fair and safe resource exchanges. Coordinate with others to determine surplus items, needed supplies, and skills that can be traded. During disruption, a community network built on cooperation is invaluable.

Support for the Vulnerable: Use your Baofeng to locate and maintain communication with those less able to search for resources themselves. The elderly, medically fragile, or those without transportation may rely on your radio to reach others for vital supplies or check-ins.

Considerations for Supply Disruptions

Sustainable Power: Prolonged shortages highlight the importance of solar chargers, hand-crank generators, and careful battery management for your Baofeng.

Community, Not Profiteering: Emphasize the importance of sharing information to alleviate shortages, not exploit them. Your Baofeng empowers fair solutions, not predatory behavior.

It's About More Than the Radio

In these complex scenarios, success rests heavily on pre-established community ties, a spirit of mutual aid, and prioritizing safety above all else. Your Baofeng is a powerful tool to reinforce these principles but doesn't operate in a vacuum.

Chapter 8: Digital Modes

Traditionally, our Baofeng radios have been tools for voice communication. But hidden within them lies the potential to transmit data, text messages, and even GPS coordinates. Digital modes open doors that simply aren't possible with analog voice alone.

Imagine your voice carried with crystal-clear perfection even when the signal is weak or overcoming noisy environments where normal conversations would be lost. Envision sending emergency supply lists, coordinating precisely with maps during disaster response, or tracking the path of a severe storm in real-time. Digital modes make this – and more – a reality.

Let's unlock the potential of these modes. We'll explore how they enhance clarity, allow us to share vital information in ways voice can't, and expand our radio network far beyond our local area. Get ready to transform your Baofeng into a powerful and adaptable communication tool!

Types of Digital Modes

Let's introduce some of the most common digital modes you might encounter:

DMR: (Digital Mobile Radio) Popular within the amateur radio community for its clear audio and advanced features like text messaging and GPS integration.

D-STAR: Another widely-used mode offering similar capabilities to DMR, along with the potential for linking repeaters to create wide-reaching networks.

APRS: (Automatic Packet Reporting System) This mode excels at sharing location data and short messages. Keep track of fellow hikers with APRS-equipped Baofengs or transmit weather readings from remote sensors.

PSK31: One of the simplest yet versatile modes designed for text-based communication. When voice transmission is unreliable, PSK31 can get your message through.

APRS Scenarios: Where it Shines

Emergency Response: During a natural disaster, when cell networks are down, APRS shines. Teams track the location of responders in real-time on maps, ensuring efficient deployment of resources. Vital supply locations, damage assessments, and even requests for assistance can be transmitted via APRS short messages, enhancing coordination in chaotic situations.

Outdoor Exploration: A group of hikers sets out, each equipped with APRS trackers. Their progress is displayed on a shared map, enabling those at basecamp to track their route and confirm their well-being. If someone ventures off-trail or requires assistance, their APRS beacon pinpoints their exact location for a quick response.

Weather Monitoring: Remote weather stations equipped with APRS transmit live wind speed, temperature, and rainfall data directly to a central map. This network of real-time data offers invaluable insights for storm preparedness, wildfire risk assessment, or even localized weather alerts for those with agricultural operations.

How APRS Works (Simplified)

The Packet Principle: Think of APRS as sending short bursts of data over radio frequencies. These "packets" contain your call sign, location (from a GPS), a short message, and potentially other information like altitude or sensor readings.

The Network Effect: Dedicated APRS receivers (often connected to the internet) pick up these packets, relaying them into a global network. Software then decodes this raw data, placing it beautifully onto maps.

Visualizing the Power: Resource Spotlight

aprs.fi: This fantastic website gives you a taste of the vast APRS network. You can zoom into your area, see live APRS activity, track moving stations, and even investigate the types of data being sent. (No radio equipment required to explore this!)

The APRS Advantage

APRS transforms your Baofeng into more than just a voice communication tool. It becomes a beacon of your position, a lifeline in emergencies, and a way to share vital data effortlessly with others. While a deeper technical dive into setting up your APRS station is needed, these scenarios illustrate the incredible potential that awaits when you harness the power of this digital mode!

Benefits of Going Digital

Clarity & Error Correction: Digital modes are engineered to handle noise and weak signals far better than traditional voice. Where an analog voice transmission might become garbled, digital can often recover the original message intact.

Information Beyond Voice: Share GPS coordinates, detailed situation reports, or crucial resource lists during emergencies where voice alone may be inefficient.

Expanding Your Network: Digital modes often have dedicated communities and integrated online infrastructure, opening new avenues for connection and knowledge exchange.

Things to Remember

Licensing Considerations: Depending on your region and specific modes, additional licensing may be required to transmit digitally. Make sure you're operating within the rules!

Learning Curve: While the advantages are compelling, digital modes frequently demand some technical learning to set up and use effectively. Patience and a spirit of experimentation are your friends.

Community is Key: Tap into online forums and local clubs with expertise in digital modes. The journey is far smoother with guidance and support!

Chapter 9: Troubleshooting and Maintenance

Think of your Baofeng as a trusty sidekick. You wouldn't let your friend run around in a bad mood without trying to help, right? Your radio deserves the same care! Regular maintenance and a knack for troubleshooting will keep it ready for your communication adventures.

The Frustration of Faulty Communication

We've all been there. You key up the mic, ready to chat, but all you hear is garbled static, your signal barely reaches the next street, or your radio seems dead. Don't throw it out the window just yet! Let's work through a systematic troubleshooting process.

Troubleshooting Basics: Start Simple

Before diving into complex theories, rule out the obvious:

Charged Up: Is the battery fully charged? A drained battery means weak transmissions, or no power at all.

Tight Connections: Check the antenna, battery, and any external speaker/mic connections. A loose connection can lead to all sorts of weird problems.

Antenna Antics: Did your antenna take a tumble, leaving it bent or damaged? Inspect it for any breaks or unusual bends.

Sometimes, the simplest fix makes you feel a bit silly afterward – we all forget the basics occasionally!

Beyond the Obvious: User Manual to the Rescue

If the simple checks don't solve it, remember that dusty user manual (who actually reads those, right?) Surprisingly, they often include a dedicated troubleshooting section. It might have the specific solution tailored to your problem.

The Power of Community: Online Forums and Videos

You're not alone! The Baofeng community is huge – enthusiasts worldwide have likely faced the same issues. Online forums, YouTube videos, and dedicated websites provide a wealth of knowledge:

Search Smart: Use specific terms describing your problem ("Baofeng UV-5R won't transmit," "Baofeng static on all channels," etc.)

Support Groups: Several dedicated Baofeng groups exist on social media platforms. Don't be shy about asking for help – you'll find others eager to share their experience.

Prevention is Key: Basic Baofeng Maintenance

A little TLC goes a long way in keeping your radio happy and healthy:

Keep it Clean: Dust and grime can work their way into connectors and buttons. A gentle wipe-down with a slightly damp cloth (never dripping wet!) removes dirt and keeps everything operating smoothly.

Connector Care: Occasionally check your antenna, battery, and external speaker/mic connectors for grime or signs of corrosion. Cleaning them with a cotton swab dipped in electronic contact cleaner can prevent intermittent connection problems.

A Case for Protection: Invest in a good case. It shields your radio from bumps, scratches, and minor drops that might damage delicate internal parts.

Water Woes: The Enemy of Electronics

Accidentally soaked your Baofeng? Don't panic, but act quickly:

Power Down: Immediately turn it off and remove the battery.

Disassemble (If Comfortable): If you're familiar with electronics, carefully open the case to allow better airflow for drying.

Avoid the Rice Trick: Rice might help with a phone, but it's less effective for complex radios. Air drying is often better.

Seek Guidance: If you're unsure, consult online resources or find a local repair technician with experience in water-damaged electronics.

The Battery Conundrum

Batteries have a lifespan. If your Baofeng suddenly loses power quickly or refuses to charge, these are the prime suspects:

Planned Obsolescence: Batteries wear out. It's unfortunate, but eventually, a replacement is necessary.

Sleepy Batteries: Letting a battery sit unused for months lowers its capacity and performance. Regularly charging your batteries keeps them healthy.

The Value of Mentors: Ham Clubs and Local Experts

Sometimes, the best troubleshooting happens hands-on. Befriend your local ham radio club, find a knowledgeable Baofeng user, or consider a reputable repair shop specializing in communication equipment. Getting experienced eyes on your problem can save you from a world of frustration.

Additional Troubleshooting Scenarios

Let's tackle a few more common scenarios:

"No Transmit" Problems: Ensure you're on the correct frequency, your PTT button actually works, and any programming restrictions (like transmit locks) aren't accidentally engaged.

Distorted Audio: Check speaker/mic connections, try a different speaker/mic to isolate the problem, and ensure there's no damage to your radio's internal speaker.

Advanced Troubleshooting Scenarios

Mysterious Menu Mayhem: Accidentally changed a setting you don't understand? Remember, most Baofeng radios have a "factory reset" option buried somewhere in the menus. Consult the manual or search online for your specific model's reset procedure. Caution: This typically erases your programmed channels, so have your channel list handy for re-entry!

Firmware Updates: Manufacturers sometimes release firmware updates that fix bugs or add features. These are typically found on brand websites. However, firmware updates are a bit advanced; only proceed if you're comfortable following manufacturer instructions precisely.

Beyond the Radio: Accessories & Environment

Sometimes the fault lies not with the Baofeng itself, but with surrounding factors:

Antenna Ailments: A broken antenna or a bad connection can severely cripple transmit and receive capabilities. If possible, temporarily try a known-good antenna to pinpoint the issue.

External Speaker Struggles: If your audio problems only occur with an external speaker, check the connector on the radio and speaker for wear or debris. Test with an alternative speaker unit when possible.

Environment Interference: While your Baofeng might excel in the wilderness, it can struggle inside buildings or areas with dense electrical noise. Try changing your location, investigate sources of possible radio frequency interference (RFI), and seek online guides about reducing RFI in your environment.

It's Not Always "Broken": Limitations & Regulations

Before declaring your Baofeng faulty, remember:

Line of Sight Matters: Baofeng radios, especially handhelds, rely heavily on unobstructed line of sight. Hills, buildings, and even dense foliage limit range more than you might expect.

Licensing & Power Restrictions Amateur radio regulations often limit transmit power and what frequencies you can legally use. Ensure you're operating within the laws of your region.

The Learning Journey: Troubleshooting as a Skill

Troubleshooting is an ongoing process. Don't get discouraged by the occasional hiccup! Consider keeping a radio notebook to track problems encountered and the solutions you find. Here's why this is valuable:

Pattern Recognition: Over time, you'll spot common issues in your setup or environment that lead to recurring problems.

Experience = Confidence: Each solved problem makes you a more knowledgeable Baofeng user, boosting your troubleshooting abilities for the future.

The Joy of Helping Others: The Baofeng community thrives on shared knowledge - your documented experience might one day help a fellow radio enthusiast!

When to Seek Professional Help

Despite your best efforts, sometimes professional expertise is needed. Here are signs it might be time to consider a qualified repair technician:

Water Woes Revisited: Severe water damage often requires specialized tools and knowledge for complete drying and potential electronic component replacement.

Hardware Malfunctions: When problems persist despite trying everything, there might be a broken component inside your radio requiring advanced diagnosis.

Warranty Considerations: If your Baofeng is new and malfunctions repeatedly, utilizing the manufacturer's warranty might be the optimal (and cost-free) path to repair.

The Well-Maintained Baofeng: Your Reliable Companion

With knowledge, patience, and a bit of preventative maintenance, you'll unlock the full capability of your Baofeng radio. Don't let minor glitches dampen your enthusiasm. Treat this as a continuous learning journey, and soon you'll be the one others turn to with their Baofeng questions!

Chapter 10: Legal and Regulatory Landscape

The world of radio transmission is a fascinating mix of freedom and responsibility. Your Baofeng empowers you to communicate far and wide, but it's crucial to be aware of the legal boundaries governing its use. This chapter will guide you through the essential regulations established by the Federal Communications Commission (FCC), ensuring you become both a skilled and ethical radio operator.

Understanding Licensing: The Foundation of Access

Amateur Radio (Ham): Obtaining a license (Technician, General, or Extra class) is your gateway to the richest experience with a Baofeng. Ham licenses grant access to a wide range of frequencies, higher power output privileges, and the opportunity to participate in the global amateur radio community.

GMRS: The General Mobile Radio Service requires a license, but no exam. It covers specific frequencies and has power limitations. Baofengs can be modified for GMRS, but always verify that your radio's modifications comply with the FCC's strict rules on this service.

Other Services (FRS, MURS, Marine): These services each have unique licensing and equipment stipulations. Before operating a Baofeng on these frequencies, thorough research into the permitted uses and modifications (if any) is crucial.

The Rules Apply, Even with a License

Having a license is fantastic, but it doesn't mean a free-for-all! Remember, as a licensed operator, you are an ambassador for the amateur radio hobby:

Call Sign Identification: The FCC mandates identifying yourself by your call sign at regular intervals (typically every 10 minutes) and at the end of transmissions. This accountability is essential for keeping the airwaves organized.

Language and Content: Obscene, disruptive, or misleading transmissions are strictly forbidden. Think of the airwaves as a shared public space; treat others with respect and avoid transmissions that could incite harmful actions.

Emergencies Prioritized: While a Baofeng is a valuable tool in crisis situations, be mindful that emergency communications take precedence. Familiarize yourself with designated emergency channels and yield them for life-or-death situations.

Power Isn't Everything: FCC Considerations

Your powerful Baofeng may have adjustable power settings, but always double-check permitted limits for the frequencies you use.

Frequency Allocations: Each radio service has power restrictions. A handy reference chart, searchable online or often provided by the ARRL, can prevent unintentional violations.

The Gray Areas: Proceed with Caution

Receive vs. Transmit: The freedom to listen is generally much broader than the freedom to transmit. Many frequencies are open for monitoring, but transmitting often needs a license.

Encryption Scrutiny: The FCC tightly regulates the scrambling of signals, particularly for certain services. Modifying your Baofeng for encryption might be explicitly illegal – always research thoroughly!

Staying Current: Your Resources

The ARRL (http://www.arrl.org): As the national association for amateur radio, the ARRL is a goldmine of information. They offer guides, legal updates, and can connect you to a network of experts.

The FCC Website (http://www.fcc.gov): This is the ultimate but sometimes dense source of truth. When the FCC changes a rule, this is where it'll be documented first.

Local Clubs and Mentors: Never underestimate the power of your local ham radio club. They're usually well-versed in regional legal nuances and have members happy to guide newcomers toward responsible radio use.

Important Disclaimer: This chapter is a starting point, and laws can change. Always treat yourself as the primary authority responsible for ensuring your Baofeng use remains legal and ethical.

Beyond the USA: International Considerations

If you travel abroad with your Baofeng, the regulatory landscape shifts dramatically! Each country has its own licensing and operating rules:

Reciprocal Agreements: Some countries offer temporary operating privileges for licensed hams from other countries. Always research the specific destination well in advance of travel.

"Listen-Only" Might Be Safest: When in doubt, listening can be a far less legally risky option than transmitting. Enjoying local radio culture is a great way to learn!

Respect Local Laws: Never assume your home country's license or permissions carry over when abroad. When traveling, being an informed and courteous radio operator is paramount.

Chapter 11: Specialized Communication Protocols

Picture this: your location automatically beaming out to a map with every step you take, text messages flying through the air even when cell signals fail, or your Baofeng chatting directly with computers... that's the magic of communication protocols!

Let's start with APRS (Automatic Packet Reporting System), the star of easily accessible data transmission. If you're into hiking, emergency preparedness, or just geeking out over live maps, APRS is your jam. We'll break down how to turn your Baofeng into a beacon, sharing your position with others and even sending short messages. Imagine plotting your entire road trip adventure with zero effort or coordinating search teams in real-time. APRS makes it happen!

Think of packet radio like the older, more tech-savvy cousin of APRS. It's all about sending chunks of digital data directly between radios. We're talking text messaging, file transfers, and even connecting your Baofeng to a whole network without any repeaters needed. Get ready to feel like a radio hacker from the '90s (in a good way)!

Don't worry, we'll demystify the gear needed to join in the fun. Things like TNCs (don't let the acronym scare you) will be explained, and we'll show you how to turn your Baofeng and a humble computer into a data-transmitting machine.

And the possibilities? Endless! Sending accurate weather data from that remote mountain cabin? Check. Email over radio waves when the internet is down? Yep, it's possible! Linking up with other hams using specialized bulletin board systems? Old-school digital fun awaits!

Fair warning: this chapter might unleash your inner tinkerer. Suddenly, the idea of sending GPS coordinates from a homemade Arduino project to your Baofeng seems totally doable... and ridiculously awesome.

Resources are your lifeline for this adventure. Get ready for websites, forums, and software recommendations to unlock the full potential of these protocols. The communities behind them are passionate and always willing to help newbies find their way.

Let's turn your Baofeng from a mere walkie-talkie into a tool for the digital age! Do you have a specific use case in mind (emergency comms, data logging, etc.)? We can tailor the examples to make this chapter even more relevant for you.

Understanding TNCs

The Bridge Builder: Think of a TNC (Terminal Node Controller) as a translator between your Baofeng and the world of digital data. It takes text, location coordinates, or other information and packages it into audio tones your radio can transmit. Conversely, it picks up those audio tones received by your Baofeng and turns them back into understandable data for your computer or other devices.

Types of TNCs: Thankfully, you have options to suit your tech comfort level and budget! Here's a basic breakdown:

Dedicated Hardware TNC: This is a small, self-contained device specifically designed for connecting to a radio. Often they have simple displays and buttons for configuration, making them a good choice for those who prefer a plug-and-play approach.

Soundcard TNC: If you're comfortable tinkering, this solution uses your computer's soundcard and freely available software to do the work of a TNC. This is generally more affordable but can involve a steeper learning curve with setup.

Built-in Capabilities: Some newer Baofeng models boast some TNC-like functionality internally. While this may be perfect for basic APRS, more advanced protocols might still require an external device for full flexibility.

The Future is Flexible: It's important to know that ongoing innovation creates exciting possibilities. New protocols and clever software solutions sometimes allow your radio to talk directly to a computer for certain tasks, bypassing the need for a traditional TNC altogether! We'll always provide up-to-date info on these options.

Choosing the Right Fit

Don't be intimidated by the tech jargon! When it's time to tackle a project, these factors will help you make the right TNC decision:

Protocol Specifics: Are you dead-set on classic packet radio or is APRS your goal? The specific communication protocol you want to use might dictate the best type of TNC.

Your Comfort Zone: Love tinkering and tweaking settings on your computer? A soundcard solution could be the perfect challenge. Prefer something ready-to-go out of the box? A dedicated hardware TNC might be your better bet.

Budget Matters: TNCs range from very affordable to those boasting advanced features at a premium price.

Your Computer: The Data Hub

Software: Your Secret Weapon: The right software breathes life into your digital communication adventures. Let's explore some popular options:

Soundcard Solutions: Versatile programs like DireWolf provide a powerful toolkit for multiple modes, including classic packet radio and APRS. These often require a bit more setup but reward you with flexibility.

APRS Focused: If Automatic Packet Reporting System is your main jam, programs like APRSIS32 or YAAC excel at decoding beacons, plotting them on maps, and making it easy to transmit your own position data.

Beyond the Basics: As we delve into specialized protocols, you'll encounter software built for tasks like weather data sharing, rig control, or even participating in digital bulletin board systems over radio waves.

Making the Connection

Diagrams to the Rescue: A simple diagram showcasing a typical setup (Baofeng -> TNC (if needed) -> Computer) will go a long way in demystifying the physical connections.

It's All About the Cables: We won't leave you guessing what kind of cables you need. Expect a clear list of common requirements (audio patch cables, serial cables if using an older TNC, etc.), including where to easily source them.

The Support Network

Resource Round-Up: I'll curate a collection of websites where you can download the best free and trial-version software to get you started.

Community Connection: Popular forums dedicated to digital modes and Baofeng use are goldmines for software support. We'll point you to the friendliest and most helpful ones for your specific software choices.

Important Note: Software choices and setup can vary slightly depending on whether you're using a traditional TNC, relying on a Baofeng's internal capabilities, or using a purely software-based solution. Rest assured, the chapter will adapt accordingly!

Chapter 12: Baofengs for Preppers and Survivalists

When the infrastructure we rely upon falters, it's the prepared who thrive. A Baofeng radio is an essential tool in the prepper's arsenal, offering a lifeline when traditional communication crumbles. Let's explore why the Baofeng's blend of affordability, versatility, and rugged simplicity make it a standout choice for survival preparedness.

Resilience over Reliance

Cellphones are marvels of modern tech, but their reliance on towers, networks, and constant power renders them fragile in a crisis. Here's where the Baofeng shines:

Battery Basics: Replaceable, rechargeable batteries ensure your radio won't become a useless brick when the grid fails. Solar charging options add an extra layer of self-sufficiency.

Durable Design: While model quality varies, many Baofengs are built tougher than sleek smartphones. Drops, water exposure, and rough handling are less likely to completely cripple your communication.

Simple and Focused: No apps, no distractions. A Baofeng's streamlined focus on voice communication conserves precious battery life and mental energy during stressful situations.

Accessing Information: Your Key Advantage

A charged Baofeng isn't just about talking; it's about unlocking crucial information:

Weather Alerts: Tap into NOAA or other weather channels for real-time updates on storms, floods, or approaching hazards affecting your survival plan.

Emergency Broadcasts: Depending on your region, there might be dedicated emergency frequencies with instructions, evacuation notices, or resource location updates.

Community Coordination: Pre-established local preparedness networks (often utilizing amateur radio frequencies) become vital conduits of information and aid organization during widespread events.

Real-World Scenarios: Where a Baofeng Excels

Let's imagine a few situations where your Baofeng becomes invaluable:

Natural Disasters: Coordinating with separated family or reaching out to neighbors in the aftermath of a storm turns chaotic without communication. A Baofeng can help locate loved ones or organize relief efforts within your community.

Remote Emergencies: Hiking or off-roading accidents leave you miles from aid. While satellite phones are ideal, a Baofeng with APRS or the ability to reach distant repeaters can still be the tool that summons help.

Grid-Down Scenarios: Extended power outages cut off landlines and cell towers. Your Baofeng aids in bartering for supplies, finding operational fuel stations, or obtaining vital updates on the restoration of critical services.

Preparation is Paramount: Beyond the Radio

Gear means little without skills. Consider these steps alongside your Baofeng investment:

Get Licensed: While emergencies might necessitate gray-area operation, a ham radio license massively expands your frequency access and allows for legal practice and participation in preparedness networks.

Join the Community: Seek out local prepper groups or ham radio clubs. Drills, knowledge sharing, and the camaraderie boost your competence and offer support networks in a real crisis.

Integrate with Your Kit: Your Baofeng isn't an island. First aid training, map and compass skills, and other elements of self-reliance turn radio communication into a force multiplier for survival.

Understanding Limitations: Responsible Use

While powerful, even a Baofeng has limits:

Line of Sight and Repeaters: Terrain and distance still matter. Explore repeaters in your area to extend range, but understand that no radio offers truly global reach without specialized satellite systems.

Etiquette and Laws: In emergencies, pragmatism may override strict rules, but always strive for respectful use of the airwaves. Interference on critical channels can have dire consequences.

Resources for Preparedness

To harness the full potential of your Baofeng as a prepper:

Dedicated Websites: Search for blogs and forums focused on preppers or survivalists who heavily utilize radio communications. Their experience is invaluable.

Offline Manuals: When the internet is down, having those critical NOAA frequencies and local channel charts pre-printed becomes crucial.

Practice Makes Perfect: Regular drills with friends or family ensure that when a real emergency strikes, using your Baofeng is second nature, not a source of fumbling stress.

Urban Survival: Communication within Chaos

Cities, with their dense infrastructure and reliance on fragile networks, pose unique survival challenges when disaster strikes. Here's how a Baofeng can make a difference:

Neighborhood Networks: During power outages and cell tower disruptions, pre-established local networks (often on GMRS or amateur radio frequencies) become arteries of information. Your Baofeng helps you check on neighbors, locate essential resources, learn evacuation routes, or perhaps organize community watch shifts in deteriorating situations.

Overcoming Obstacles: Urban environments are rife with signal-blocking structures. Experiment beforehand to discover "sweet spots" near windows or higher floors where you get better reception. Pre-programming your Baofeng with crucial frequencies (local emergency channels, police/fire bands if legal to monitor, etc.) saves precious time in a crisis.

Family Coordination: Getting separated during evacuations or chaotic events is a major concern. A simple, pre-decided channel and call-in times with family members equipped with Baofengs can provide peace of mind or become a lifeline for reunification.

Monitoring the "Big Picture": Tuning into news channels or frequencies used by relief agencies might provide wider-scope updates unobtainable through neighborhood networks. This broader awareness could alert you to approaching dangers, resource distribution locations, or shifts in the overall situation.

Wilderness Survival: A Voice in the Vastness

Remote terrain and the absence of infrastructure place a premium on any method of signaling and communication. Your Baofeng becomes your potential means to summon aid or share critical information:

The Power of Repeaters: Learn about repeaters in your region, especially those on hilltops or tall structures. With a bit of power and the right frequency, your Baofeng might reach far enough to catch their attention, vastly extending your ability to call for help in a remote emergency.

APRS and Location Sharing: If equipped with the capabilities, APRS (Automatic Packet Reporting System) allows sending GPS coordinates via radio. Search and rescue teams might monitor these frequencies, giving them your precise location even if a voice call is impossible.

Group Coordination: Hiking or hunting parties separated by miles can maintain contact with well-chosen frequencies and regular check-ins. This can mitigate the danger of individuals becoming lost or injured unnoticed by their companions.

Weather Alerts: Remote areas often lack reliable weather information. Monitoring NOAA or dedicated weather channels through your Baofeng provides early warning about storms or sudden shifts in conditions that could turn a trip dangerous.

Additional Considerations (Urban and Wilderness)

Power and Batteries: Never rely on a single battery! Solar chargers, hand-crank options, or abundant spares are crucial. In both environments, a low-power setting extends operating time when maximum range isn't strictly necessary.

Signal Enhancement: Even a basic handheld Baofeng can be improved. Higher-gain antennas, especially if temporarily elevated (think of attaching it atop a tall hiking pole), often make the difference between a weak signal and getting through.

Discretion and Security: In some survival situations, broadcasting your location or activities on an easily monitored radio might be unwise. Learn the encryption limitations of your Baofeng, and decide if it warrants pre-decided code words or low-profile communication plans.

The Importance of Practice: Before Crisis Strikes

In both urban and wilderness survival, the time to learn your radio is NOT during an emergency. Here's how to ensure your Baofeng is a tool you know well:

Familiarize yourself with Settings: Changing frequencies, power output, and squelch should be second-nature, not a source of fumbling stress when it matters.

Local Terrain Matters: Test your Baofeng's limits in your area. Know those dead zones in your neighborhood, or how far your signal reaches reliably before terrain blocks it in the wilderness.

Participate in Drills: Community preparedness groups and ham radio clubs often run simulated emergency drills – these are invaluable learning experiences!

Chapter 13: Ham Radio Operations with Baofengs

The world of ham radio, also known as amateur radio, bustles with a vast community of enthusiasts. They connect across towns, states, and even countries using radio waves. Obtaining an official amateur radio license unlocks a whole new dimension of what you can do with your Baofeng.

Let's dispel a common misconception: while emergencies are one focus of ham radio, it's about far more! Hams explore the science of radio, experiment with building their own gear, participate in friendly competitions, and some even chat with astronauts on the International Space Station. It's a hobby where the limits are truly what you make of them.

So, why should you get licensed? Firstly, it grants you access to a wider range of frequencies specifically set aside for ham radio use. This means less crowded airwaves and the potential for longer-distance communication. Secondly, you'll gain the knowledge and respect for proper radio etiquette, crucial when sharing frequencies with others. While using your Baofeng within certain rules without a license might be necessary in a true emergency, think of getting licensed as upgrading your preparedness toolkit for routine practice and the full benefit of being part of the community.

Your Baofeng can be your first step on this journey. Many models possess features that, when paired with the right license, allow you to participate in voice conversations on local repeaters (imagine them as powerful relays boosting your signal), experiment with digital modes, and much more. The first step is finding a license class that interests you. In many countries, there are entry-level licenses focusing on the basics and granting access to a useful range of frequencies.

Ham radio is a fantastic social hobby! Seek out local clubs or online communities dedicated to helping newcomers. They often run classes for license exams, provide hands-on experience with different radios, and are always happy to answer all your Baofeng and ham radio questions.

Resources are your best friend in this journey. Websites like the ARRL (http://www.arrl.org) in the US or similar national organizations in your country are treasure troves of information for getting licensed and discovering all that ham radio offers. Don't be afraid to ask for help – the ham radio community is known for its welcoming spirit to those eager to learn.

Remember: there's always more to explore in the world of ham radio. From tinkering with antennas to chasing faraway signals as they bounce off the atmosphere, the possibilities are endless. Your Baofeng, paired with a license, is your ticket into this incredible world of communication, experimentation, and lifelong learning.

The Experience Factor: Voices from the Airwaves

Sometimes the best way to understand the possibilities a ham radio license offers is to hear from those who've taken the plunge. Sarah, a self-proclaimed "non-techie" mom, always kept a Baofeng in her emergency kit but felt there was more it could do. After completing a beginner's license class and nervously taking that first exam, her world changed. "Suddenly, instead of just monitoring weather alerts, I was chatting with people across town, helping with communications for my kid's Scouting Jamboree... it made my Baofeng feel ten times more powerful!"

For Jason, an avid outdoorsman, getting licensed was about reliable communication in the backcountry. "Cell signal is a myth where I like to hike. My Baofeng and ham license meant I could check in with my wife, access trail condition updates from park rangers, and even hit a repeater to chat with folks miles away while resting on a mountain peak. It's a different level of connectedness."

These stories illustrate the transformation getting licensed brings. Suddenly the Baofeng isn't merely a tool for emergencies but a gateway to everyday adventures, community involvement, and expanding your communication horizons beyond anything you imagined.

A Day in the Life of a New Ham

Picture yourself as a newly licensed ham. You fire up your trusty Baofeng, program in the frequency of your local repeater, and nervously press the transmit button. A friendly voice welcomes you, and soon you're discussing the weekend weather forecast with someone down the street. Later that day, you participate in a special event hosted by your ham club, logging contacts with other enthusiasts from across the region. That evening you experiment with a digital mode, sending a text-like message that bounces across hundreds of miles. The possibilities feel exhilarating!

Addressing Hesitations: It's Easier Than You Think

The idea of any exam can bring back memories of stressful classrooms and intimidating textbooks. However, the amateur radio licensing process is far more approachable than many realize. Forget needing an engineering degree or mastering complex formulas. Think of it like learning the essential "rules of the road" for operating respectfully and safely on the airwaves.

Don't let fears about the exam hold you back! A wealth of resources exist to help you succeed. Online communities offer practice tests to gauge your knowledge and study guides that explain radio concepts in easy-to-understand language. Many local ham clubs even run dedicated classes for newcomers, providing a supportive environment to tackle the material and get hands-on experience.

The ham radio community thrives on mentorship. Seasoned hams are incredibly passionate about sharing their knowledge and are always happy to answer your questions, offer guidance, and help you get comfortable with the idea of getting licensed.

"Tech Isn't My Thing" Reassurance

You don't need to be an electronics expert to thrive as a ham! The beauty of amateur radio lies in its diversity. Some hams are fascinated by the technical side, building antennas and experimenting with their own gear. Others are drawn to the sheer joy of conversation, chatting with people from all walks of life across the globe.

If all the technical jargon makes you break out in a cold sweat, don't worry! Start with the basics – learn how to use your Baofeng and master proper operating procedures. As your confidence grows, you'll find yourself naturally absorbing knowledge from fellow hams and online resources. Dive into areas of the hobby that pique your interest and don't be afraid to experiment. Enthusiasm and a willingness to learn are far more important than being a tech whiz on day one.

Remember, getting licensed isn't just about emergencies. It's about unlocking a whole new world of communication and connection through your Baofeng. Imagine making friends across the country or tinkering with homemade antennas – the possibilities are endless!

Practical Benefits Beyond Emergencies

While preparedness is vital, ham radio weaves itself into the fun and adventure of daily life. Imagine a group hike where cell coverage is non-existent. With just a few Baofengs and your amateur radio license, you maintain easy contact with your dispersed group. No more worrying about stragglers or shouting yourself hoarse to communicate over distances. An overlanding expedition becomes infinitely safer when you can coordinate navigation with the rest of your convoy even where cell towers are mere myths.

The possibilities don't stop at voice communication. Imagine using your Baofeng to send your GPS coordinates with APRS on that solo hike, providing peace of mind for those at home. Got a question about a trail closure? Tap into local repeaters and the knowledge of fellow hams who frequent the area.

Skywarn Spotters: When the Weather Turns

If your region experiences severe weather threats, getting licensed opens the door to becoming part of the Skywarn program (or a similar network in your country). These dedicated volunteers provide real-time weather observations directly to the National Weather Service. Hams act as the eyes and ears on the ground, reporting hail, tornadoes, and critical measurements with the accuracy cell phone apps can't achieve. This kind of weather reporting is both incredibly rewarding and offers a tangible way your radio skills directly benefit your entire community.

The Adventure Starts Here

Enjoying these practical benefits hinges on getting licensed and expanding your communication knowledge. Seek out resources and training specific to hiking and outdoor use of ham radio. Learn how to participate in organized nets for events, discover which frequencies in your area are geared towards outdoor enthusiasts, and potentially connect with other Baofeng-wielding hikers or overlanders.

Beyond the Baofeng: A Taste of Progression

While your trusty Baofeng is a fantastic gateway to amateur radio, it represents just the beginning of your journey. As you gain experience and discover the activities that excite you most, a whole world of specialized ham radio equipment opens up.

Perhaps you'll become fascinated by antennas, experimenting with building your own and maximizing signal efficiency for specific tasks. Maybe the lure of higher power transmission beckons, allowing you to make reliable contacts over even greater distances with dedicated mobile or home station rigs. Digital modes pique your curiosity, leading you towards radios specifically designed to push the boundaries of data communication.

The beautiful thing about ham radio is that the pathways are endless. Some hams dedicate themselves to chasing far-away contacts under challenging conditions, testing their skills to see just how far their signal can reach. Others find joy in the technical side, designing and building equipment with their own hands. Participating in competitions or "field days" adds an element of friendly rivalry and skill-building. There's truly something for everyone!

Don't think of outgrowing your Baofeng as a negative – it's the sign of a growing passion! Your humble radio will always have value, whether as a dependable backup, an introduction tool to lend to curious friends, or a reminder of how far you've come in this rewarding hobby.

Call to Action: Find Your Tribe

The US ham radio community is vast and incredibly welcoming to those eager to learn. Here's how to connect and find your place:

Your Local Network: The ARRL (http://www.arrl.org/find-a-club) website is your golden ticket to finding active ham clubs in your area. These are treasure troves of hands-on experience, mentorship, and often even run license classes. Don't be shy about reaching out – they love welcoming newcomers!

The Online World: Forums like those on [invalid URL removed] ([invalid URL removed]) and dedicated subreddits (search for "amateur radio" or "ham radio") are bustling with fellow hams at all experience levels. Get your questions answered, share your progress, and find a virtual community to support you.

License Class Resources: The Knowledge Quest Begins

Let's empower you to ace that exam! Here are some top resources for US-based license prep:

The ARRL: Head to their "Getting Licensed" section (http://www.arrl.org/getting-licensed) for official study guides tailored to each license class, practice exams, and listings of instructors near you.

Ham Radio Prep: This website (https://www.hamradioprep.com/) offers comprehensive online courses, making the learning process convenient and structured.

Tried and True: The book "Ham Radio License Manual" published by the ARRL is considered a classic study resource by many.

Search Power: Don't underestimate searches like "ham radio license class [Your State]" for finding local groups offering exam prep or online courses with a regional focus.

You're Not Alone

The journey from Baofeng enthusiast to licensed ham is incredibly rewarding. Tap into the collective knowledge of the community, seek mentorship, and don't be afraid to ask questions. Your callsign and the world of ham radio await!

Chapter 14: Unleashing the Potential

While your Baofeng is a capable communication tool right out of the box, it's also a blank canvas begging for experimentation and customization. Let's explore ways to push it beyond its factory limits and tailor it to your unique needs.

Important Disclaimer: Modifying your radio can be risky! Always proceed with caution, research thoroughly, and consider the potential impact on your radio's warranty and compliance with local regulations.

Upgrades for the Beginner: Easy Wins, Big Impact

Antennas Rule: One of the simplest, most effective upgrades is the antenna. Aftermarket options abound, from longer whips for better reach to specialized antennas for specific bands. Play with placement too – even a few feet higher can transform your signal.

Power Play: Batteries are the lifeline of portable radios. Extended capacity options keep you on the air longer during emergencies or outdoor adventures. Solar chargers, hand-cranked backups, or clever car battery adapters provide diverse power solutions for any scenario.

Software Secrets: CHIRP software is your gateway to deeper, menu-less customization. It lets you add or delete channels easily, fine-tune power settings, and tailor radio behaviors. CHIRP's compatibility extends beyond Baofeng, making it a great tool to have as your radio interest grows.

The Next Level: When Simple Isn't Enough

As your needs become more specialized, hardware modifications pave the way to exciting new possibilities:

Loud and Clear: In noisy environments, an external speaker drastically improves audio. If you're monitoring in your vehicle or operating near machinery, this mod can turn garbled transmissions into clear communication.

Filtering for Focus: Radio environments can be cluttered with stray signals. Adding tailored filters to your Baofeng helps it lock onto your desired frequency, improving reception clarity.

Digital Adventures: Ready to go beyond voice? Hardware interfaces let you utilize your Baofeng for data modes like APRS (sharing GPS location) or PSK31 (text-based communication). This requires more technical knowledge, but greatly expands possibilities!

The DIY Spirit: Embracing the Journey

The heart of the ham radio community is experimentation and knowledge sharing. Seek out resources that spark your tinkering spirit:

Baofeng Mod Guides: Websites like https://miklor.com/ offer dedicated Baofeng projects tailored to different skill levels. You might be surprised how approachable some of these enhancements are.

Forums for Fellow Tinkerers: Online communities are full of users who've walked the same path. Swap tips, troubleshooting advice, and discover the sheer range of possibilities.

Reach out to Your Local Ham Club: Nothing beats hands-on guidance. Ham radio clubs often have 'tech nights' where experienced members help others with projects.

The Rewards: Empowerment and Personalization

Whether you just add a bigger battery or venture into advanced modifications, the act of tailoring your Baofeng is satisfying and empowering. It's a process that deepens your understanding of radio fundamentals, teaches practical skills, and leaves you with a tool uniquely tuned to the way YOU want to communicate.

Weatherproofing: Shield Your Radio from the Elements

If your adventures often involve rain, snow, or dusty conditions, basic weatherproofing can make a world of difference:

Seal the Seams: Silicone sealant, carefully applied along any casing gaps or around buttons, creates a barrier against water and dust intrusion. Research your specific Baofeng model for areas to focus on.

Conformal Coating: For the advanced DIYer, conformal coating sprayed over the internal electronics offers superior moisture protection. Disassembly is required, so proceed only if you're comfortable.

Protective Cases: Often inexpensive, a simple case adds a layer of shock and water resistance. Choose ones with seals around buttons, speaker, and mic ports.

The Low-Tech Option: Even a well-sealed Ziplock bag in a pinch can save your Baofeng from sudden downpours. Consider this a "last resort" for emergencies rather than a permanent solution.

Data Modes: Expanding Communication Horizons

Unlocking digital modes on a budget-friendly Baofeng opens doors to new ways of sharing information:

APRS: Location and More: With specialized hardware, your Baofeng can transmit GPS coordinates over amateur frequencies. This is invaluable for search-and-rescue, outdoor tracking, and even participating in events like balloon launches.

PSK-31: Text Over the Airwaves: PSK-31 uses sound to transmit text. With a computer interface, your Baofeng becomes a tool for sending emails or relaying messages when voice communication is difficult.

Interfaces: The Bridge: Data modes require specialized cables or devices to interface your Baofeng with a computer or GPS receiver. Projects like Mobilinkd TNCs or DIY solutions based on soundcards bridge the gap.

Digital Communities: Seek out forums and groups specializing in digital ham radio modes. The learning curve can be steeper, but you'll find experienced users eager to guide newcomers.

Beyond the Basics: Niche Mods for Every Need

The world of Baofeng modification is vast. Here are some more niche ideas to pique your curiosity:

Frequency Extension: Unlocks Restricted Bands: Some Baofengs can be modified for transmitting outside designated amateur bands (proceed with EXTREME caution regarding legality!).

External Mic/PTT Upgrades: For mobile or base station setups, custom setups with handheld mics or footswitch PTT controls can improve operating comfort.

Battery Overhauls: Repacking your Baofeng battery with higher-capacity cells is possible for the technically inclined, boosting your on-air time significantly.

Important Reminders

Skill vs. Risk: Always weigh the potential benefits of a modification against your comfort level and the risk of damaging your radio.

Regulations Rule: Never forget that the responsibility to follow your local radio laws rests entirely with you, even after modifying your equipment.

Community is Key: The best source of targeted info is often other Baofeng users who've done similar projects. Tap into the knowledge pool!

Upgrade 1: Magnetic Mount Antenna for Mobile/Temporary Use

Benefits:

Dramatic Signal Boost: A large, vehicle-mounted antenna vastly improves range.

Temporary Setup: No drilling holes or permanent changes to your vehicle.

Flexibility: Move it between cars or for base station use (on a metal surface).

Materials:

Magnetic Mount Antenna: Select one designed for your frequency bands. Popular options include dual-band VHF/UHF antennas or those tailored for specific frequencies.

Coaxial Cable: High-quality, low-loss cable is crucial. Ensure it's long enough to comfortably reach from your antenna location to your Baofeng inside the vehicle.

Correct Connectors: Your cable must have the right type (PL-259, SMA, etc.) to match both the antenna and your radio.

Steps:

Choose Antenna Location: Ideally, the center of your vehicle's roof offers the best signal spread. However, trunk lids or side panels can also work.

Clean the Surface: Ensure the mounting area is free of dirt or debris that might scratch your vehicle or weaken the magnet's hold.

Position the Antenna: Gently place the mount with antenna attached. Test reception with your Baofeng before fully tightening down the mount (if it has adjustment features).

Route the Cable: Carefully run the cable into your vehicle. Avoid pinch points like doors, and keep it away from moving parts (seat mechanisms).

Connect and Tune: Securely attach the coax to your Baofeng. Use the SWR meter function (if your radio has one) or an external meter to fine-tune the antenna for the lowest SWR reading on your desired frequencies.

Upgrade 2: Improving the Handheld Antenna

Benefits:

Inexpensive: A simple antenna swap can often provide a noticeable boost.

Beginner Friendly: No complex wiring or disassembly is required.

Variety: Many options exist, from longer whips to specialized antennas.

Materials:

Aftermarket Antenna: Choose an antenna compatible with your Baofeng's connector type (SMA-Female is most common) and designed for your operating frequencies. Consider portability vs. performance gain when choosing the length.

Steps:

Remove Stock Antenna: Carefully unscrew the original "rubber ducky" antenna.

Install New Antenna: Gently but firmly screw on your new antenna, ensuring a snug connection.

Test it Out! Tune into a known active frequency or transmit to a friend for a signal report. Compare the results to your previous antenna.

Important Notes:

Connector Care: Always hand-tighten connections, never use pliers, to avoid damage.

Weather Matters: Outdoor antennas may have specific grounding needs for lightning protection.

SWR Checks: Especially for vehicle mounts, checking SWR ensures your antenna is performing optimally and prevents potential damage to your radio.

Resources:

YouTube Search: Baofeng Magnetic Mount: [invalid URL removed] – find visual guides for various setups.

https://miklor.com/: https://miklor.com/ – Search for articles on antennas and mounts.

Local Ham Clubs: Often have equipment demos or members with hands-on experience.

Benefits of an External Speaker

Overcoming Noise: In loud environments (vehicles, outdoors with machinery nearby, etc.), a good external speaker cuts through the background noise.

Enhanced Monitoring: When you need to monitor a frequency for long durations, the improved audio reduces fatigue and lets you focus.

Accessibility: A well-positioned speaker can make it easier to hear for those with hearing impairments.

Materials

Small Powered Speaker: Choose one with volume control and ideally 8 Ohms impedance to match the Baofeng. Options range from dedicated ham radio gear to repurposed computer speakers.

Audio Jack: A 3.5mm jack matching your desired speaker. Be aware many Baofengs have a mono audio output, even with a stereo jack on the radio itself.

Wire & Soldering Tools: Insulated wire (speaker cable works well) and basic soldering iron, solder, and flux.

Heat Shrink Tubing: To neatly insulate your connections.

Steps:

Disassembly: Power off and remove the battery from your Baofeng. Depending on your model, this will likely involve removing a few screws and carefully separating the case halves.

Locate the Speaker Points: Consult online resources like https://miklor.com/ or model-specific forums to identify the internal speaker solder points. These are usually small metal pads directly connected to the tiny speaker.

Careful Soldering: Prepare your wires (strip ends, pre-tin with solder). Very carefully, solder one wire to each of the speaker pads. Avoid applying too much heat or touching other components.

External Connection: Route the wires out of the Baofeng's casing. Ensure it can reassemble without pinching them. Attach the wires to your 3.5mm jack, paying attention to polarity (+/-).

Insulate and Test: Use heat shrink tubing over your solder joints and the exposed jack. Before closing the Baofeng, plug in the speaker and carefully power on for a test. Correct sound output confirms your connections are right.

Reassembly and Placement: Close up your Baofeng. Consider creative ways to mount your external speaker in your car, at your base station, etc. Velcro, adhesive strips, or custom brackets can enhance usability and prevent damage.

Important Notes:

Disassembly at Your Own Risk: Internal work voids warranties and risks damaging your radio. Proceed only if you're comfortable with small electronics.

Polarity Matters: Incorrect wiring to your jack can damage the speaker or radio. Double-check before testing.

Volume Control: If your speaker lacks volume control, consider adding a small in-line volume knob to the wire.

Alternatives: Some Baofengs have dedicated external speaker ports – check your model's documentation, simplifying this mod greatly.

Resources

Photo Guides: Search for "<Your Baofeng Model> External Speaker Mod" – visual guides are invaluable.

Forums: If stuck, online communities often have users who've done this on your exact radio model.

Leveling Up:

For further audio upgrades, explore adding filters (DIY or pre-built modules) inside your Baofeng to reduce noise on received frequencies. This requires more advanced soldering skills but can improve clarity significantly.

Conclusion: The Future of Baofeng and Your Communication Toolkit

Your journey with the Baofeng radio doesn't end with the final page of this guide. The beauty of this versatile device lies in its ability to grow alongside your interests. Perhaps you started as a prepper focused on emergency readiness but discovered a passion for storm spotting through your ham radio license. Maybe your initial forays into antenna building sparked a desire to explore the world of satellite communication.

Technology continues to evolve, and with it, so does the Baofeng. Newer models boast improved features, wider compatibility, and push the boundaries of what we expect from affordable handheld radios. Keep a curious eye on these advancements as they may add even more possibilities to your radio adventures.

The real heart of your communication toolkit isn't just the Baofeng itself, but rather the knowledge and skills you acquire throughout your exploration. Learning about frequencies, mastering operating procedures, understanding radio wave propagation – these are assets that carry over regardless of the specific gear you might utilize.

View your Baofeng as both a gateway and a constant companion. It ignited your interest in the wider world of radio, served as a training ground for new skills, and reliably powered your on-air experiences. As your needs become more sophisticated, you may graduate to specialized equipment, but your trusty Baofeng will always hold value as a backup, an introductory tool for others, or a symbol of where your communication journey began.

The world of radio is one of constant discovery and connection. May your Baofeng guide you towards exciting encounters, enduring friendships, and the knowledge that no matter the circumstance, you have the power to be heard.

BONUS 1: Essential Reference: Frequencies, Phonetics, and More

This chapter serves as your on-the-go guide for those crucial radio knowledge bits you'll find yourself using again and again. Consider it a toolbox for navigating frequencies in the United States, ensuring clear communication, and maximizing your Baofeng's potential.

United States Frequency Allocations

Understanding the vast world of radio frequencies can be overwhelming. Federal Communications Commission (FCC) regulations govern specific uses and licensing requirements. Always double-check the FCC website (https://www.fcc.gov/) for the latest information before transmitting.

Emergency Services:

Police/Fire/EMS: Monitoring these frequencies may be restricted in some states. Check your local regulations before listening. National Emergency Channels:

GMRS Channel 12 (462.550 MHz): Designated for emergency use by the FCC.

NOAA Weather Radio (See "Weather & Information" below): Often used for emergency broadcasts.

Weather & Information:

NOAA Weather Radio: A vital resource for real-time weather updates, severe weather warnings, and forecasts. Locate your closest station's frequency using the National Weather Service's searchable map: https://www.weather.gov/nwr/station_listing

Traveler Information Stations (varies by location): Often found on major highways, these stations provide traffic updates and road condition information.

Ham Radio Bands:

The world of ham radio opens up as you progress through your license levels. Here's a quick breakdown of commonly used bands and resources for further exploration:

Technician Class (Tech):

2 Meter Band (144 - 148 MHz, Tx/Rx): A popular entry point for hams, offering local communication and repeater access.

70 Centimeter Band (420 - 450 MHz, Tx/Rx): Another popular band for local and regional communication.

General Class (General): In addition to Technician privileges, General licenses unlock access to:

High Frequency (HF) Bands (various ranges, Tx/Rx): These bands allow communication over vast distances depending on conditions.

Extra Class: The pinnacle of amateur radio licensing grants access to the full range of frequencies allocated for ham radio use.

Finding Repeaters: Repeaters are strategically placed stations that amplify your signal, allowing you to communicate over longer distances. Websites like RepeaterBook (https://www.repeaterbook.com/repeaters/index.php) can help you locate repeaters in your area.

Beyond the Basics:

The world of radio communication extends far beyond these listed frequencies. Here are a few additional options to explore:

GMRS/FRS: These license-free services offer short-distance communication options. Be aware of power limitations and channel sharing etiquette.

MURS (Multiple Use Radio Service): Another license-free option with specific channel allocations.

Local Event/Group Channels: Some parks, events, or volunteer organizations may utilize designated frequencies for coordination purposes.

Section 2: The Phonetic Alphabet

Clear communication is essential in any radio environment. The phonetic alphabet, also known as the NATO phonetic alphabet, ensures clarity when transmitting letters or spelling out callsigns. Here's the table for reference:

A = ALFA

B = BRAVO

C = CHARLIE

D = DELTA

E = ECO

F = FOXTROT

G = GOLF

H = ALBERGO

I = INDIA

J = GIULIETTA

K = CHILOL=LIMA

M = MIKE

N = NOVEMBRE

O = OSCAR

P = PAPA

Q = QUEBEC

R = ROMEO

S = SIERRA

T = TANGO

U = UNIFORME

V = VINCITORE

W = WHISKY

X = RAGGI X

Y = YANKEE

Z = ZULU

Bonus 2: Upgrade Paths for Your Baofeng

Your Baofeng isn't just a radio; it's a starting point for a fulfilling communication journey. Here's how to strategically evolve your setup over time, keeping both your budget and specific interests in mind:

Tier 1: The Immediate Essentials

Batteries and Power: A high-capacity rechargeable battery (or several spares) eliminates limitations for extended use. If off-grid operation is a possibility, a small solar panel adds incredible flexibility.

Antenna Improvement: Research a better aftermarket antenna for the bands you use most frequently. Even a slightly longer "whip" can make a big difference. Resources like YouTube often show real-world range comparisons.

Comfort and Protection: A simple case protects your investment, while a speaker/mic tailored to your operating style (in the car, at home, etc.) enhances usability and audio clarity.

Tier 2: Expanding Your Horizons

Software Exploration: CHIRP lets you easily manage and organize even complex channel lists. Start experimenting with custom power settings and less common options in your Baofeng's menus for finer control.

Beyond Voice: If interested in digital modes, research a basic interface for APRS or PSK31. Start with listening and beaconing (APRS) before moving on to full two-way exchanges.

Station Enhancement: Is your Baofeng primarily a home base? Invest in a desktop power supply that replaces finicky batteries. A dedicated external speaker improves monitoring over long periods.

Tier 3: Specialized and Advanced

Niche Antennas: Do you frequent specific repeaters with known locations? A directional antenna pointed their way dramatically boosts your signal.

Mobile Mastery: For serious vehicle use, consider professional SWR tuning of a magnetic mount antenna, proper grounding, and perhaps even a dedicated 12V power distribution system for a clean, reliable setup.

Filtering and Beyond: If strong local interference plagues your reception, an internal filter mod (requires soldering skill) or an external, band-specific filter can transform a muddled signal into a clear one.

The Path is Your Own: A Few Guiding Principles

Needs vs. Wants: Do you mostly use one band around town, or are you a frequency explorer? Upgrade based on how you ACTUALLY use the radio, not just on cool-sounding features.

Skill and Comfort: Love tinkering? Mods like external speakers are approachable. Never touched a soldering iron? Focus on accessories and software.

Learning as You Go: The process of researching and implementing upgrades teaches you more about radio than just using it. This knowledge is invaluable!

Community is Key: Stuck on a choice? Ham radio thrives on shared knowledge. Local clubs or online forums help you decide the BEST upgrade for your situation.

Emergency Preparedness: When Reliability is Paramount

Focus Area: Self-sufficiency, extended use without infrastructure, local information

Power Supremacy: Multiple large-capacity batteries are a must. Solar charging, even a small panel, becomes incredibly valuable. Consider hand-crank or dynamo chargers for true worst-case scenarios.

Knowing the Frequencies: Pre-program your Baofeng with emergency services, weather channels (NOAA), and any dedicated regional disaster response nets. CHIRP software makes this easy.

Neighborhood Network: If available, focus on GMRS compatibility (licensing required!) alongside your amateur radio bands. This facilitates local community coordination when cell towers fail.

The Printed Advantage: Have a hard copy of important frequencies and procedures alongside your Baofeng. When you're stressed or the internet is down, paper can be a lifesaver.

Outdoor Adventures: Communication Off the Grid

Focus Area: Distance, portability, maximizing location advantages

Antenna Power: A high-gain roll-up antenna can be packed away, then deployed on a tree branch for a temporary range boost. Learn how to safely elevate antennas in the field.

APRS Mastery: If allowed in your area, an APRS interface turns your Baofeng into a beacon. Share your position with fellow hikers or search and rescue teams in remote areas.

Simplicity as Backup: A set of pre-decided simplex frequencies for hiking groups eliminates reliance on repeaters that might be out of reach. Practice this beforehand!

Weather is King: Regular NOAA weather checks are vital when venturing off the beaten path. An external speaker makes monitoring easy at camp or while on the move.

City Living: Overcoming Signal Barriers

Focus Area: Cutting through urban noise, finding hidden repeaters, reliable home operations

Seek the High Ground: Experiment with where you get the best reception in your home. A Baofeng near a window, or its antenna run outdoors, can be a game-changer.

Invest in Filtering: Dense urban environments are rife with radio-frequency interference. A simple external band-pass filter can work wonders in cleaning up reception.

Repeater Research: Use online resources like RepeaterBook to map repeaters in your city. Program these, even weaker ones, as you may be surprised what you reach with a little elevation.

Home Base Power: A plug-in power supply and external speaker upgrade a Baofeng to a desktop monitoring station, perfect for catching local nets or being at the ready.

Universal Upgrade Reminders

A few principles apply no matter how you use your Baofeng:

Practice Matters: Run drills with friends or family, get to know your radio well before you truly need it.

Protection and Portability: A basic case and a well-organized bag for accessories keeps your Baofeng safe and ready to go.

Community Knowledge: Local clubs or online groups help you make informed upgrade choices and discover what works in YOUR specific environment.

www.ingramcontent.com/pod-product-compliance
Lightning Source LLC
Chambersburg PA
CBHW062226220526
45471CB00009B/3363